SpringerBriefs in Materials

Series Editors

Sujata K. Bhatia, University of Delaware, Newark, USA

Alain Diebold, Schenectady, USA

Juejun Hu, Department of Materials Science and Engineering, Massachusetts Institute of Technology, Cambridge, USA

Kannan M. Krishnan, University of Washington, Seattle, USA

Dario Narducci, Department of Materials Science, University of Milano Bicocca, Milano, Italy

Suprakas Sinha Ray , Centre for Nanostructures Materials, Council for Scientific and Industrial Research, Brummeria, Pretoria, South Africa

Gerhard Wilde, Altenberge, Germany

The SpringerBriefs Series in Materials presents highly relevant, concise monographs on a wide range of topics covering fundamental advances and new applications in the field. Areas of interest include topical information on innovative, structural and functional materials and composites as well as fundamental principles, physical properties, materials theory and design.

Indexed in Scopus (2022).

SpringerBriefs present succinct summaries of cutting-edge research and practical applications across a wide spectrum of fields. Featuring compact volumes of 50 to 125 pages, the series covers a range of content from professional to academic. Typical topics might include

- A timely report of state-of-the art analytical techniques
- A bridge between new research results, as published in journal articles, and a contextual literature review
- A snapshot of a hot or emerging topic
- An in-depth case study or clinical example
- A presentation of core concepts that students must understand in order to make independent contributions

Briefs are characterized by fast, global electronic dissemination, standard publishing contracts, standardized manuscript preparation and formatting guidelines, and expedited production schedules.

Jia Li · Nor Azmira Salleh · Norariza Ahmad ·
Adil Alshoaibi · Soorathep Kheawhom ·
Ahmad Azmin Mohamad

Reduced Graphene Oxide for Supercapacitor Applications

Jia Li
Energy Materials Research Group (EMRG)
School of Materials and Mineral Resources
Engineering
Universiti Sains Malaysia
Nibong Tebal, Pulau Pinang, Malaysia

Norariza Ahmad
Advanced Materials and Subsea
Technology
PETRONAS Research Sdn Bhd
Kajang, Selangor, Malaysia

Soorathep Kheawhom
Department of Chemical Engineering
Faculty of Engineering
Chulalongkorn University
Bangkok, Thailand

Nor Azmira Salleh
Energy Materials Research Group (EMRG)
School of Materials and Mineral Resources
Engineering
Universiti Sains Malaysia
Nibong Tebal, Pulau Pinang, Malaysia

Adil Alshoaibi
Department of Physics
College of Science
King Faisal University
Al Hofuf, Saudi Arabia

Ahmad Azmin Mohamad
Energy Materials Research Group (EMRG)
School of Materials and Mineral Resources
Engineering
Universiti Sains Malaysia
Nibong Tebal, Pulau Pinang, Malaysia

ISSN 2192-1091   ISSN 2192-1105   (electronic)
SpringerBriefs in Materials
ISBN 978-981-96-4929-7   ISBN 978-981-96-4930-3   (eBook)
https://doi.org/10.1007/978-981-96-4930-3

© The Editor(s) (if applicable) and The Author(s), under exclusive license to Springer Nature
Singapore Pte Ltd. 2025

This work is subject to copyright. All rights are solely and exclusively licensed by the Publisher, whether the whole or part of the material is concerned, specifically the rights of translation, reprinting, reuse of illustrations, recitation, broadcasting, reproduction on microfilms or in any other physical way, and transmission or information storage and retrieval, electronic adaptation, computer software, or by similar or dissimilar methodology now known or hereafter developed.
The use of general descriptive names, registered names, trademarks, service marks, etc. in this publication does not imply, even in the absence of a specific statement, that such names are exempt from the relevant protective laws and regulations and therefore free for general use.
The publisher, the authors and the editors are safe to assume that the advice and information in this book are believed to be true and accurate at the date of publication. Neither the publisher nor the authors or the editors give a warranty, expressed or implied, with respect to the material contained herein or for any errors or omissions that may have been made. The publisher remains neutral with regard to jurisdictional claims in published maps and institutional affiliations.

This Springer imprint is published by the registered company Springer Nature Singapore Pte Ltd.
The registered company address is: 152 Beach Road, #21-01/04 Gateway East, Singapore 189721, Singapore

If disposing of this product, please recycle the paper.

Preface

Since its first isolated and identified in 2004, graphene has attracted much attention in the scientific community due to its excellent properties. The excellent conductivity, single-layer carbon structure, extremely high strength, and high transparency of graphene are highly respected in academic circles. However, due to its zero-bandgap characteristics and the complexity of large-scale production, graphene faces a series of challenges in direct applications.

To address these challenges, reduced graphene oxide (rGO) offers a practical solution as a chemically modified derivative of graphene. RGO is obtained by reducing graphene oxide (GO), which is the oxidized form of graphene. This reduction process not only restores the SP^2 carbon hybrid structure of graphene, but also introduces hydrogen (-H) or carbon (-C) functional groups, which significantly improves the conductivity and other key properties of graphene. Compared with pure GO, rGO has more diversity. The transition from graphene to rGO represents a major shift from basic research to practical applications. This chemical modification makes the material more suitable for a variety of applications and provides a key technical basis for practical applications of graphene in electronics, materials science, and other fields. This process marks an important step forward from laboratory research to commercialization and industrialization of graphene materials.

This book describes in detail several common methods for the preparation of rGO, including chemical, biological, photoreduction, and thermal reduction methods, each of which has its own unique advantages and challenges. Chemical reduction can effectively restore the conductivity of GO, but impurities may be introduced, and biological reduction has the advantage of environmental protection, but its reduction efficiency is relatively low; heat treatment is simple and can significantly improve the electrochemical properties of rGO, but high temperature may lead to material damage; the purity of thermal reduction is high, but it needs special light source and condition. Through the comprehensive analysis of these methods, readers can fully understand the advantages and disadvantages of various methods and provide guidance for the selection of appropriate methods in practical application.

The application of rGO in electrode materials for supercapacitors is introduced in detail in this book. The characteristics of rGO electrode are analyzed by physical and electrochemical characterization, and its advantages including high capacitance, excellent charge-discharge rate, and long-term stability are discussed. The use of rGO materials can significantly improve the performance of supercapacitors and enhance their energy storage capacity and power output, so that it has a wide range of applications in the field of energy storage.

The book is intended to fill a gap in current research on the practical applications of graphene and its derivatives; it aims to provide a more comprehensive and detailed reference for researchers, engineers, students, and others interested in the field. It systematically outlines:

- Unique properties of graphene and rGO.
- Currently used methods for synthesizing rGO.
- Preparation techniques and applications of rGO films.
- Exceptional performance of rGO films.
- Performance analysis of rGO films as electrode materials for supercapacitors.

The book is divided into six chapters. Chapter 1 introduces the fundamentals of graphene, GO, and rGO. Chapter 2 focuses on the synthesis methods of rGO. Chapter 3 describes the preparation techniques and properties of rGO films. Chapter 4 provides a detailed analysis of the physical properties of rGO. Chapters 5 and 6 respectively concentrate on cyclic voltammetry (CV) and galvanostatic charge-discharge (GCD) analyses, examining the electrochemical characteristics of rGO.

Nibong Tebal, Malaysia	Jia Li
Nibong Tebal, Malaysia	Nor Azmira Salleh
Kajang, Malaysia	Norariza Ahmad
Al Hofuf, Saudi Arabia	Adil Alshoaibi
Bangkok, Thailand	Soorathep Kheawhom
Nibong Tebal, Malaysia	Ahmad Azmin Mohamad

Acknowledgements This work was supported by the Fundamental Research Grant Scheme (FRGS) (FRGS/1/2024/TK09/USM/02/7), Ministry of Higher Education, Malaysia. This work was supported by the Deanship of Scientific Research, Vice Presidency for Graduate Studies and Scientific Research, King Faisal University, Saudi Arabia [Grant No. KFU242590].

Competing Interests The authors have no competing interests to declare that are relevant to the content of this manuscript.

Contents

1 **Introduction to Reduced Graphene Oxide** 1
   1.1 Introduction 1
   1.2 Type of Graphene 2
       1.2.1 Graphene 2
       1.2.2 Graphene Oxide 3
       1.2.3 Reduced Graphene Oxide 4
   1.3 Structure and Properties of Reduced Graphene Oxide 5
   1.4 Applications of Reduced Graphene Oxide 5
       1.4.1 Energy Storage 6
       1.4.2 Sensors 6
       1.4.3 Supercapacitors 6
       1.4.4 Biomedical Applications 8
   1.5 Challenges and Future Perspectives 9
   1.6 Summary 9
   References 10

2 **Synthesis of Reduced Graphene Oxide** 13
   2.1 Introduction 13
   2.2 Synthesis of Graphite-Graphene 14
   2.3 Synthesis of Graphite-Graphene Oxide 15
   2.4 Synthesis of Graphene Oxide—Reduced Graphene Oxide 15
       2.4.1 Chemical Methods 16
       2.4.2 Biological Pathways 18
       2.4.3 Photoreduction 19
       2.4.4 Thermal Reduction 20
   2.5 Summary 23
   References 23

3 **Reduced Graphene Oxide Thin Film Electrode** 27
   3.1 Introduction 27
   3.2 Preparation Method of Reduced Graphene Oxide Film 28

		3.2.1	Spin Coating	29
		3.2.2	Vacuum Filtration	29
		3.2.3	Electrostatic Spraying	30
		3.2.4	Dip Coating	31
		3.2.5	Drop Casting	31
		3.2.6	Electrophoresis	32
	3.3	Application of Reduced Graphene Oxide Film		33
		3.3.1	Energy Storage	33
		3.3.2	Sensor	34
		3.3.3	Flexible Electronics and Smart Materials	36
	3.4	Current Challenges and Future Directions		36
		3.4.1	Quality Control	36
		3.4.2	Large-Scale Production	36
		3.4.3	Innovative Preparation Technology	37
		3.4.4	Performance Optimization	38
	3.5	Summary		38
	References			38
4	**Structural, Morphology, and Chemical Species Properties of Reduced Graphene Oxide**			**41**
	4.1	Introduction		41
	4.2	Structure Analysis		42
	4.3	Morphologies Analysis		44
	4.4	Chemical Species Analysis		46
	4.5	Summary		49
	References			50
5	**Cyclic Voltammetry Analysis of Reduced Graphene Oxide for Supercapacitors**			**53**
	5.1	Introduction		53
	5.2	Experimental Setup for Cyclic Voltammetry		54
	5.3	Calculation for Cyclic Voltammetry		55
	5.4	Cyclic Voltammetry of Graphene Oxide and Reduced Graphene Oxide		57
	5.5	Effect of Scan Rate on Cyclic Voltammetry		58
	5.6	Effect of Composite Materials on Cyclic Voltammetry		60
	5.7	Summary		62
	References			62
6	**Galvanostatic Charge–Discharge Analysis of Reduced Graphene Oxide for Supercapacitors**			**65**
	6.1	Introduction		65
	6.2	Charge–Discharge Fabrication for Reduced Graphene Oxide Electrodes		66

6.3	Charge–Discharge Calculation	66
6.4	Effect of Reduction Time on Galvanostatic Charge–Discharge	67
6.5	Effect of Current Density on Galvanostatic Charge–Discharge	69
6.6	Effect of Nanocomposite Materials on Galvanostatic Charge–Discharge	71
6.7	Summary	72
References		73

Chapter 1
Introduction to Reduced Graphene Oxide

Abstract Reduced graphene oxide (rGO) is a derivative of graphene that retains a similar structure but includes various oxygen-containing functional groups and structural defects. These characteristics result in superior electrical conductivity, excellent mechanical properties, and enhanced chemical stability. RGO demonstrates advantages in its cost-effective synthesis process, providing a large specific surface area for chemical reactions and superior chemical stability compared to both graphene and graphene oxide. These properties make rGO highly potential in various energy system technologies, including energy storage, sensors, and supercapacitors.

Keywords Graphene · Graphene oxide · Reduced graphene oxide · Supercapacitor

1.1 Introduction

Graphene is discovered by Andre Geim and Konstantin Novoselov in 2004 [1, 2]. It is a monolayer of carbon atoms arranged in a honeycomb-hexagonal structure [3], giving it excellent chemical and mechanical properties with wide applications in energy, sensors, and biomedical fields [1, 2, 4]. Graphene oxide (GO) is formed by the introduction of a large number of oxygen-containing functional groups [5], which oxidize graphene, resulting in products typically existing in powder form [6]. RGO is created by introducing reducing agents into the GO structure, thereby partially restoring the structure and properties of graphene [7].

This chapter discusses the structure with respective properties of graphene, GO and rGO in details to highlight the unique properties of rGO in the applications of energy storage, sensors, supercapacitors, and biomedical. This chapter also explained the challenges remained and also discussed the directions for further exploration on the potential of rGO in different areas.

1.2 Type of Graphene

1.2.1 Graphene

Graphene is a two-dimensional (2D) carbon allotrope, composed of a single layer of carbon atoms arranged in a unique honeycomb lattice (Fig. 1.1a) [1, 4]. This arrangement not only gives it unique physical and chemical properties but also makes it particularly eye-catching in the field of materials science. Graphene is unique in that it can be viewed as both a strong solid material and a giant molecule with an astonishing molecular weight of 10^6 to 10^7 g mol^{-1} [8]. In natural graphite, graphene layers are tightly bonded through π-π stacking interactions. This is represented graphically as one plane of layered graphite separated from the bulk crystal, and this non-covalent interlayer bonding provides graphite with extremely high thermodynamic stability (Fig. 1.1b) [8–10].

Graphene has a high specific surface area (2630 m^2 g^{-1}) [11], extraordinary strength (Young's modulus 1.1 Tpa, tensile strength 125 GPa) [12], higher electron conductivity than graphite (electron mobility of 150,000 cm^2 V^{-1} s^{-1}) [13], and thermal conductivity (thermal conductivity of 5300 W m^{-1} K^{-1}) [14]. These properties, combined with its low density and flexibility, make graphene an ideal material for applications such as supercapacitors. Its immense specific surface area and minimal density enable efficient electron–ion interactions and rapid electron migration rates [10, 15]. This enhances the electrical conductivity and capacitance characteristics, thereby increasing the charge storage necessary for high-performance capacitors. However, in the multilayer structures of graphene, due to the strong van der Waals interactions between individual layers, research in the wet chemistry of graphene

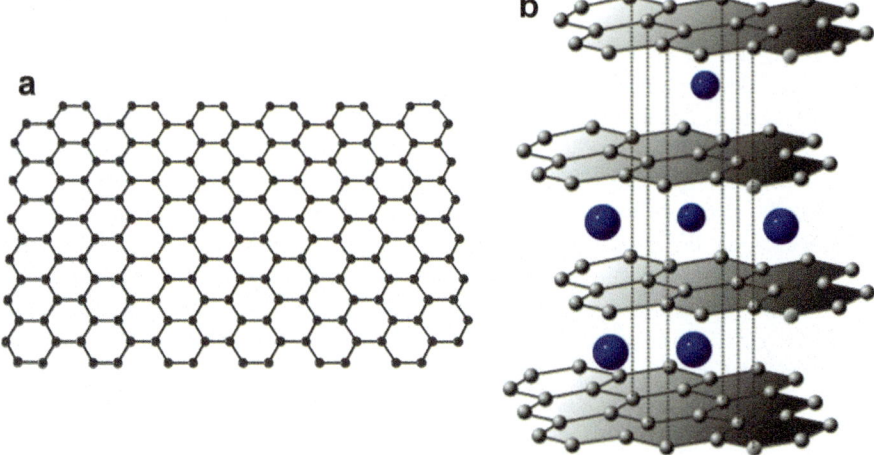

Fig. 1.1 **a** Structure diagram of graphene and **b** simplified structure of graphite layers, reproduced with permission from Ref. [8]

1.2.2 Graphene Oxide

Graphene oxide (GO) is a two-dimensional carbon-based nanomaterial derived from graphene (Fig. 1.2a) [1], whose unique properties have become a focal point of scientific research in recent years [16]. GO is a derivative of graphene obtained through chemical oxidation processes (often employing the Hummers method) [9, 17]. This oxidation effectively mitigates the challenges associated with wet chemical processing of graphene. The method involves oxidizing natural graphite using strong oxidants such as concentrated sulfuric acid, nitric acid, and potassium permanganate, to introduce oxygen-containing functional groups like hydroxyl (–OH), epoxy (–O–), and carboxyl (–COOH) onto the graphene layers, thereby transforming it into GO (Fig. 1.2b) [8, 9, 17].

The structural characteristics of GO differentiate it from graphene in several ways. Firstly, due to the introduction of oxygen-containing functional groups in GO, part of the sp^2 hybrid is transformed into sp^3 hybrid, which changes the hexagonal structure inherent in traditional graphene; this affects the chemical and electronic properties of its surface [4, 8, 10, 15, 18]. Especially, its hydrophilicity and polarity are enhanced significantly, which makes it have a wide application prospect in the fields of energy storage and electronic facilities [4].

Secondly, the oxidation process of graphene also fundamentally changes the electronic structure of graphene, from a conductive material to a semiconductor or insulation properties of a completely new material [1, 8]. In addition, the oxidation process increases the interlayer spacing of graphene lamellae and reduces π-π stacking interactions, making it more suitable for the separation and treatment of single-layer or few-layer graphene [8, 17, 19]. This oxidation process not only changed the chemical and physical properties of graphene, enhanced its compatibility with other compounds, but also significantly affected its electronic structure. Its applications

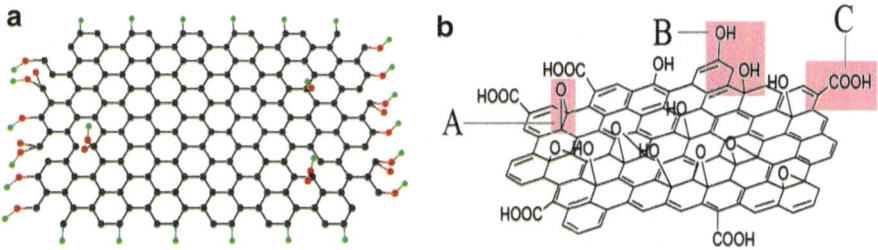

Fig. 1.2 Graphene oxide **a** atomic structure diagram and **b** chemical structure diagram, reproduced with permission from Ref. [8]

in fields such as electronic devices and energy storage systems have opened up new possibilities [18].

1.2.3 Reduced Graphene Oxide

Reduced graphene oxide (rGO) is produced through chemical, thermal, or electrochemical reduction methods (Fig. 1.3a) [1]. This process involves the partial or complete removal of oxygen functional groups such as hydroxyl (–OH), carboxyl (–COOH), and epoxy (–O–) from GO (Fig. 1.3b) [20]. It effectively mitigates the challenges faced in the wet chemical processing of graphene, thereby restoring some of its characteristics, especially electrical conductivity [19].

In the oxidation process, the existence of oxygen-containing functional groups destroyed the sp^2 hybrid structure of graphene, resulting in a significant reduction in its conductivity [19]. During the preparation of rGO, some of these oxygen functional groups are removed, partially restoring the sp^2 hybridized structure and the electron π-π conjugated system of graphene, which aids in enhancing its electrical conductivity [8, 19]. Compared to GO, rGO although containing fewer oxygen functional groups retains enough oxygen functional groups to maintain good dispersion in water and other polar solvents, making it easier to manipulate during wet chemical treatment [18]. In addition, rGO has excellent thermal stability; even at the extreme temperature of 800 °C, its total mass loss is only 11%. Because of its low production cost and excellent performance, rGO has a wide application prospect in many fields, such as energy storage, sensors, super capacitors, and so on [1, 21].

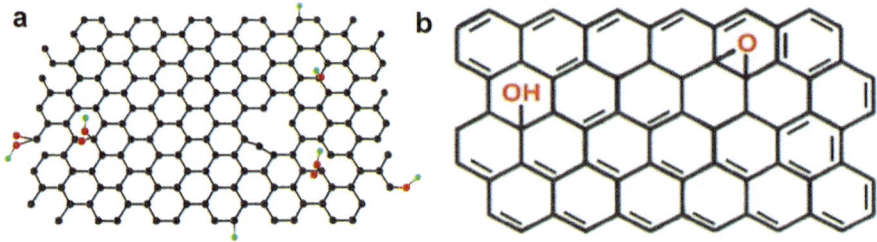

Fig. 1.3 Reduced graphene oxide **a** atomic structure diagram and **b** chemical structure diagram, reproduced with permission from Ref. [20]

1.3 Structure and Properties of Reduced Graphene Oxide

The rGO has become a hot research area because of its excellent optical, electrical, mechanical, and thermal stability [22]. RGO has high conductivity (6300 S cm^{-1}) [21] and high mobility (320 cm^2 V^{-1} s^{-1}), and its lamellae have remarkable mechanical strength, Young's modulus is about (1.0 TPa) [23], and fracture strength is about (130 GPa) [24]. The structure of rGO is similar to that of original graphene, but its unique feature is that there are a few oxygen functional groups such as epoxy group (C_2H_4O), carboxyl group (COOH), hydroxyl group (OH), and ketone group (C=O). The existence of these functional groups makes rGO have unique physical and chemical characteristics, which make it widely used in various fields [1].

1.4 Applications of Reduced Graphene Oxide

As a derivative of graphene, rGO retains many excellent properties of graphene and is easier to process and functionalize. With its high electrical conductivity, large specific surface area, and high mechanical strength, rGO shows broad application prospects in energy storage, sensors, supercapacitors, and biomedicine (Fig. 1.4).

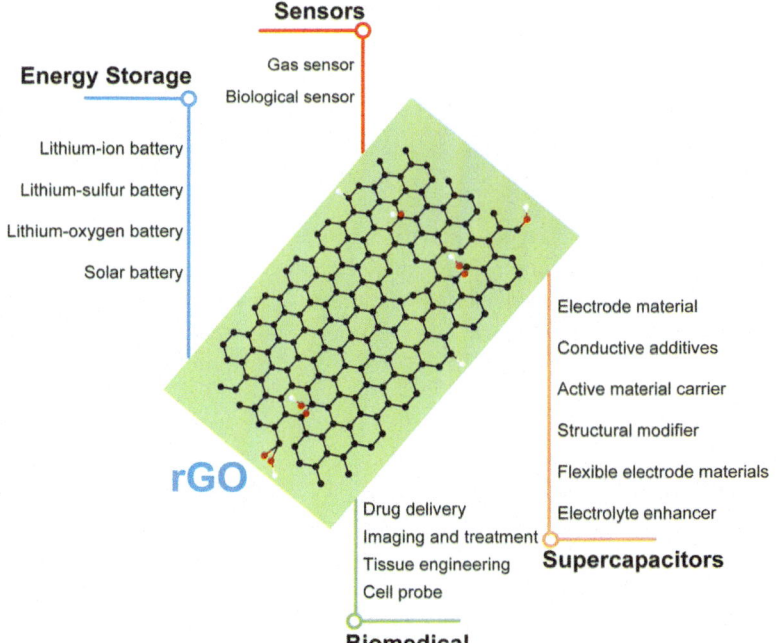

Fig. 1.4 Applications of reduced graphene oxide

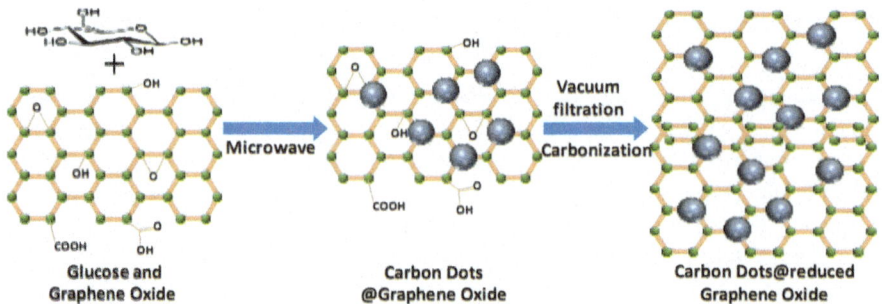

Fig. 1.5 Schematic illustration of the fabrication process for the freestanding Carbon Dots@reduced graphene oxide (CDs@rGO) paper, reproduced with permission from Ref. [25]

1.4.1 Energy Storage

With its high specific surface area and high conductivity, rGO is widely utilized as an electrode material for various rechargeable batteries, such as lithium-ion batteries, lithium-sulfur batteries, and lithium-oxygen batteries (Fig. 1.4). In a study exploring the enhancement of lithium-ion battery conductivity, carbon dots are anchored onto CDs@rGO, aiming to increase active sites and improve structural stability (Fig. 1.5) [25]. It is shown that the CDs@rGO paper anode material can maintain a high capacity of 310 mAh g^{-1} and a long cycle life at a current density of 100 mA g^{-1}, and a capacity of 244 mAh g^{-1} after 840 cycles at a high current density of 200 mA g^{-1}.

1.4.2 Sensors

The rGO material shows great potential for high-performance sensor applications due to its high specific surface area, excellent electrical conductivity, and good biocompatibility (Fig. 1.4). In a study of explosives detection, researchers use rGO to develop a unique sensor. The results show that the sensor has excellent selectivity and can effectively distinguish DNT from other substances such as acetone and ethanol [26].

1.4.3 Supercapacitors

The excellent performance of rGO material enhances the high efficiency and long life of supercapacitors. Excellent conductivity is the key to excellent performance, which is critical for high-performance energy storage and fast charge–discharge cycles. Its large specific surface area enables it to store more charge, which further increases the energy density of the supercapacitor, while its chemical stability ensures its reliability under various environmental conditions (Fig. 1.4). A modified polyhedral

1.4 Applications of Reduced Graphene Oxide

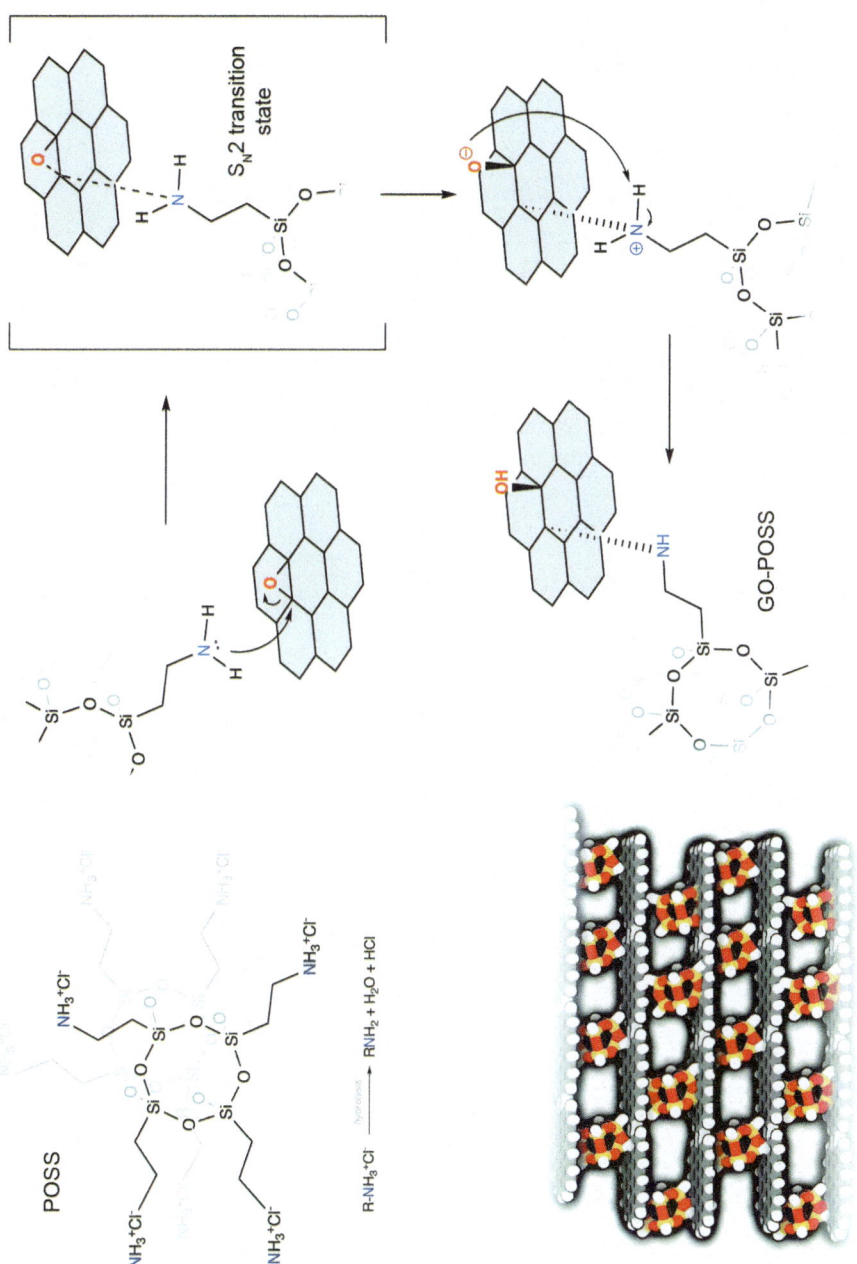

Fig. 1.6 Synthesis process of rGO-POSS hybrid, reproduced with permission from Ref. [27]

oligomeric silsesquioxane (POSS) is used to enhance the electrochemical properties of rGO. RGO-POSS hybrid materials exhibit high specific capacitance (174 F g^{-1}), significant power density (2.25 W cm^{-3}), and high energy density (41.4 mW h cm^{-3}). In addition, these electrode materials exhibit excellent durability and retain more than 98% of their capacity after 5000 cycles (Fig. 1.6) [27].

1.4.4 Biomedical Applications

The rGO materials are widely used in biomedical applications due to their unique properties such as high conductivity and strong interaction with a variety of molecules (such as drugs, proteins or other biomolecules); this gives it immense application potential in the field of biomedicine (Fig. 1.4). In a study on cancer therapy, an rGO-based gene delivery system is developed utilizing rGO functionalized with octaarginine and anti-HER2 antibodies for siRNA delivery. This system showcases the feasibility of gene silencing in cancer treatment (Fig. 1.7) [28].

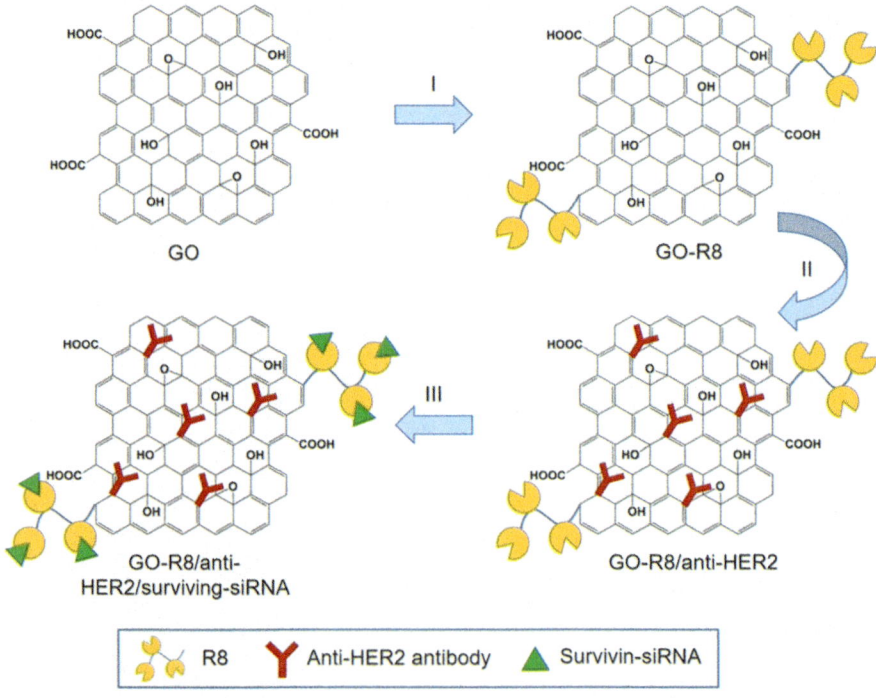

Fig. 1.7 Schematic illustration of graphene oxide functionalization with octaarginine and anti-HER2 antibody, reproduced with permission from Ref. [28]

1.5 Challenges and Future Perspectives

Technical challenges represent a key obstacle in the research and application of rGO. Despite widespread interest in rGO due to its exceptional electrochemical properties and lower production costs, its scale-up production still faces significant challenges [29]. Currently, the availability of efficient and environmentally friendly methods for rGO production is limited, constraining its widespread adoption in commercial and industrial applications. For instance, the most commonly used chemical reduction methods employ reductants that are often toxic; improper handling of these substances could result in severe environmental damage. Therefore, the development of novel and eco-friendly reduction pathways is one of the hotspots in current research [30, 31].

Functionalization and stability control represent another major challenge in the utilization of rGO. The surface functional groups of rGO significantly influence its physicochemical properties, and the qualitative customization of material properties can be achieved through precise control over the types and distribution of these functional groups. However, the technology of precise chemical modification is not mature yet and needs further development. In addition, the stability of rGO under different conditions is very important for its practical application, and further improvement of its stability is still needed [32].

Looking to the future, rGO is moving toward more efficient, more environmental protection, and more diversity direction. With the development of various new methods and technologies, the application range of rGO will be further expanded [33]. In the field of energy storage, more efficient batteries and supercapacitors can be made by improving the conductivity and surface chemical property of rGO [1, 15, 27]. In the field of biomedicine, rGO biocompatibility has been modified to develop new drug delivery systems and biosensors [31, 34].

1.6 Summary

This chapter provides a detailed introduction to rGO. As a derivative of graphene, rGO possesses unique structural defects and various oxygen functional groups. It systematically outlines:

(i) As an important derivative of graphene, rGO has many excellent properties and unique chemical properties of graphene.
(ii) With its excellent electrical conductivity and mechanical strength, rGO demonstrates extensive prospects for applications in energy storage, sensors, supercapacitors, and biomedical fields.
(iii) The multifunctional microstructures of rGO play a key role in its wide application, which shows its important position in the future development of technology.

References

1. Ahmed A et al (2022) Synthesis techniques and advances in sensing applications of reduced graphene oxide (rGO) composites: a review. Compos Part A: Appl Sci Manuf 107373
2. Geim AK, Novoselov KS (2007) The rise of graphene 6(3):183–191
3. Li S et al (2021) Highly stable two-dimensional metal-carbon monolayer with interpenetrating honeycomb structures. 5(1):52
4. De Padova P et al (2022) Defects in two-dimensional elemental materials beyond graphene. Defects in two-dimensional materials. Elsevier, pp 43–88
5. Nanjundappa V et al (2023) Efficient strategies to produce graphene and functionalized graphene materials: a review. 14:100386
6. Chen J et al (2021) One step electrochemical exfoliation of natural graphite flakes into graphene oxide for polybenzimidazole composite membranes giving enhanced performance in high temperature fuel cells. 491:229550
7. Feng J et al (2020) Synthetic routes of the reduced graphene oxide. 74:3767–3783
8. Nebol'Sin V, Galstyan V, Silina Y (2020) Graphene oxide and its chemical nature: multi-stage interactions between the oxygen and graphene. Surf Interfaces 21:100763
9. Deshwal N et al (2023) A review on recent advancements on removal of harmful metal/metal ions using graphene oxide: experimental and theoretical approaches. Sci Total Environ 858:159672
10. Lu X et al (2023) Polymer-based solid-state electrolytes for high-energy-density lithium-ion batteries–review. 13(38):2301746
11. Zhi D et al (2021) A review of three-dimensional graphene-based aerogels: synthesis, structure and application for microwave absorption. Compos B Eng 211:108642
12. Xiao W et al (2023) Three dimensional graphene composites: preparation, morphology and their multi-functional applications. Compos A Appl Sci Manuf 165:107335
13. Choi MS et al (2021) High carrier mobility in graphene doped using a monolayer of tungsten oxyselenide. Nat Electron 4(10):731–739
14. Zhao H-Y et al (2022) Efficient preconstruction of three-dimensional graphene networks for thermally conductive polymer composites. Nano-Micro Lett 14(1):129
15. Fang B et al (2020) A review on graphene fibers: expectations, advances, and prospects. Adv Mater 32(5):1902664
16. Tarcan R et al (2020) Reduced graphene oxide today. J Mater Chem C 8(4):1198–1224
17. Esencan Turkaslan B, Filiz Aydin M (2020) Optimizing parameters of graphene derivatives synthesis by modified improved Hummers. Math Methods App Sci
18. Trikkaliotis DG et al (2021) Graphene oxide synthesis, properties and characterization techniques: a comprehensive review. ChemEngineering 5(3):64
19. Gao B et al (2020) Preparation of single-layer graphene based on a wet chemical synthesis route and the effect on electrochemical properties by double layering surface functional groups to modify graphene oxide. Electrochim Acta 361:137053
20. Tadyszak K, Wychowaniec JK, Litowczenko JJN (2018) Biomedical applications of graphene-based structures. 8(11):944
21. Ramesh P et al (2023) Green approach for the synthesis of monolayer reduced graphene oxide: one-step protocol with simultaneous iodination and reduction. New J Chem
22. Razaq A et al (2022) Review on graphene-, graphene oxide-, reduced graphene oxide-based flexible composites: from fabrication to applications. Materials 15(3):1012
23. Wan S et al (2021) High-strength scalable graphene sheets by freezing stretch-induced alignment. Nat Mater 20(5):624–631
24. Abazari S, Shamsipur A, Bakhsheshi-Rad HR (2021) Reduced graphene oxide (RGO) reinforced Mg biocomposites for use as orthopedic applications: mechanical properties, cytocompatibility and antibacterial activity. J Magnes Alloy
25. Zhang E et al (2020) Carbon dots@ rGO paper as freestanding and flexible potassium-ion batteries anode. Adv Sci 7(15):2000470

26. Li Y et al (2020) Flexible TPU strain sensors with tunable sensitivity and stretchability by coupling AgNWs with rGO. J Mater Chem C 8(12):4040–4048
27. Czepa W et al (2020) Reduced graphene oxide–silsesquioxane hybrid as a novel supercapacitor electrode. Nanoscale 12(36):18733–18741
28. Bellier N et al (2022) Recent biomedical advancements in graphene oxide-and reduced graphene oxide-based nanocomposite nanocarriers. Biomater Res 26(1):65
29. Yang S et al (2020) Emerging 2D materials produced via electrochemistry. Adv Mater 32(10):1907857
30. Manikandan V, Lee NY (2023) Reduced graphene oxide: biofabrication and environmental applications. Chemosphere 311:136934
31. Mohd Kaus NH et al (2021) Effective strategies, mechanisms, and photocatalytic efficiency of semiconductor nanomaterials incorporating rGO for environmental contaminant degradation. Catalysts 11(3):302
32. Ahmed A et al (2023) Synthesis techniques and advances in sensing applications of reduced graphene oxide (rGO) composites: a review. Compos Part A: Appl Sci Manufact 165:107373
33. Destiarti L et al (2023) Challenges of using natural extracts as green reducing agents in the synthesis of rGO: a brief review. Res Chem 101270
34. Daniyal M, Liu B, Wang W (2020) Comprehensive review on graphene oxide for use in drug delivery system. Curr Med Chem 27(22):3665–3685

Chapter 2
Synthesis of Reduced Graphene Oxide

Abstract As a derivative of graphene, reduced graphene oxide (rGO) retains many of the excellent properties of the original graphene, while its honeycomb sp^2 network structure enhances charge separation and transport, resulting in high electron mobility, superior electrical conductivity, and a large specific surface area. By detailing various reduction methods for rGO, including chemical, biological, thermal reduction, and photoreduction techniques, the study analyzes their advantages, application ranges, and effects on rGO performance. The study reveals the specific effects of various reduction methods on the quality, morphology, and electrochemical performance of rGO. Through detailed analysis of the synthesis and application of various reduction methods, the extensive applications of rGO in energy storage, sensors, and supercapacitors are highlighted.

Keywords Graphene · Graphene oxide · Reduced graphene oxide · Reduction methods · Application fields

2.1 Introduction

Since its discovery in 2004, graphene has attracted significant research and commercial interest due to its applications in various industries [1]. When used directly, graphene can cause phase separation in composites due to π-π interactions and hydrophobic-hydrophobic interactions between its layers [2, 3]. As a derivative of graphene, rGO restores the intrinsic properties of graphene and is highly regarded for its excellent thermal stability and broad application prospects [4, 5]. This chapter will provide a detailed analysis of the exfoliation of graphite to form graphene, the oxidation to form graphene oxide, and the various methods for reducing graphene oxide (GO) to rGO. The reduction step plays a crucial role in the synthesis of rGO and has a significant impact on the properties of the rGO product [4, 6, 7].

A variety of reduction methods have been developed, including chemical, biological, thermal, and photo-assisted techniques [8]. These methods aim to effectively remove oxygen-containing functional groups from GO, thereby restoring its conductivity to levels similar to that of the original graphene [9–11]. This chapter discusses and analyzes the reaction conditions, reduction mechanisms and specific applications of various reduction techniques in detail. The choice of reduction method will significantly affect the morphology, structure and electrochemical properties of rGO [4, 12], so it is crucial to select an appropriate reduction method for a specific application.

This chapter introduces the synthesis process from graphite to rGO in detail, analyzes the various reduction methods used in the synthesis of rGO, and studies the unique properties of various reduction methods and their wide applications in electronics, energy storage, and sensors.

2.2 Synthesis of Graphite-Graphene

At present, there are two kinds of methods for preparing graphene: top-down and bottom-up (Fig. 2.1) [13–15]. Bottom-up means that some small carbon-containing compounds "Grow" into graphene, high-quality graphene can be prepared by chemical vapor deposition and epitaxy methods, but the production is low and the cost is high, difficult to mass produce [16]. Top-down methods involve exfoliating larger graphite particles into thin-layer graphene using external forces [17]. Techniques such as ball milling and chemical oxidation–reduction are used for the mass production of graphene, but the resulting products are of lower quality and have more defects. Each method has its unique advantages and application scenarios and has made significant contributions to graphene research.

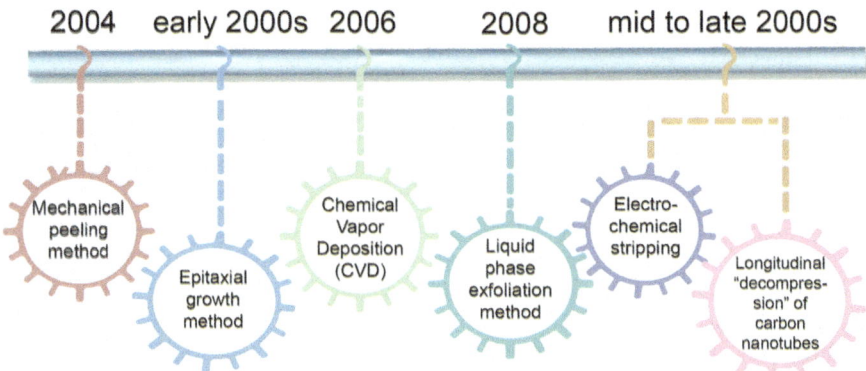

Fig. 2.1 Development history of graphene synthesis method

2.3 Synthesis of Graphite-Graphene Oxide

Graphite is one of the main sources to synthesis GO. Generally, there are three steps to synthesize GO from graphite [18]:

(i) Chemical agents or mechanical means are usually used to diffuse and exfoliate graphite.
(ii) Exfoliated graphite sheets are exposed to strong oxidants to introduce oxygen-containing functional groups on the surface of the graphite sheets.
(iii) Ultrasonic treatment to form single or few-layer GO.

The modified Hummers method is most commonly used to prepare GO, which involves oxidizing graphite in which the traditional potassium chlorate ($KCLO_3$) is replaced with potassium permanganate ($KMnO_4$) during the oxidation stage [19]. During the preparation process, a mixture of sulfuric acid (H_2SO_4) and $KMnO_4$ is used as the main oxidant, while potassium persulfate ($K_2S_2O_8$) is used to maintain a stable pH value in the reaction environment [20, 21]. In addition, in order to reduce the generation of acid mist, sodium nitrate ($NaNO_3$) is used instead of nitric acid (HNO_3). By adjusting the ratio of $KMnO_4$ and H_2SO_4, not only the yield of GO is improved, but also the generation of toxic gases (NO_2 and N_2O_4) is effectively reduced [22, 23]. Due to its unique properties, GO prepared by this method has broad application potential in many fields such as composite materials, transparent conductive films, solar energy, and biomedicine.

2.4 Synthesis of Graphene Oxide—Reduced Graphene Oxide

The preparation of rGO involves the removal of oxygen-containing functional groups in GO, a process that can be achieved through a variety of approaches, including chemical, biological, thermal treatment, and light-assisted methods (Fig. 2.2) [24]. Each method may lead to differences in the morphology and electrical properties of rGO. In the reduction design of GO, key elements include precise control of the C/O ratio, selective removal of target oxygen groups (hydroxyl, carboxylate, epoxy groups), repair of GO surface defects, and the use of green reduction agents [25], as well as maintaining or improving the physical and chemical properties of GO (mechanical strength, electrical conductivity, optical properties, and dispersion) [26].

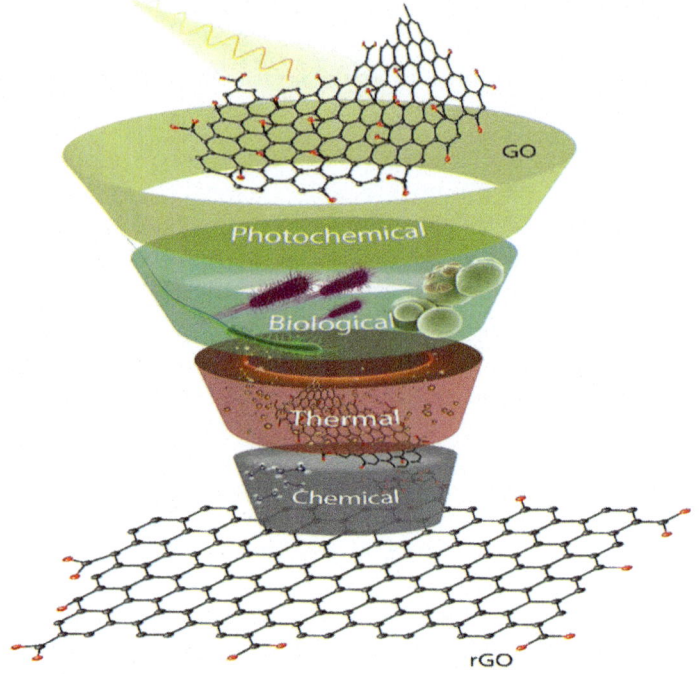

Fig. 2.2 Various techniques for the synthesis of rGO, reproduced with permission from Ref. [24]

2.4.1 Chemical Methods

Hydrazine

Hydrazine monohydrate is the most widely used reducing agent for reducing GO. Using hydrazine monohydrate as a reducing agent, the oxygen-containing functional groups in GO can be effectively removed, restoring its structure and giving it electronic and structural properties similar to original graphene. The process of reducing exfoliated graphene oxide sheets with hydrazine in water produced a material, rGO, with graphitic properties similar to the original graphite; the resulting rGO has a conductivity of 2×10^2 S m^{-1} [19]. Studies have shown that by heating GO at 95 °C for 3 h, the atomic ratio of carbon to oxygen can be increased from 3.1 to 15.1. This result reflects the significant effects of temperature and time on the properties of rGO [27].

Catalyst

In the reduction of GO, aluminum (Al) and zinc (Zn) are the most commonly used catalysts, which can complete the highly efficient reduction reaction in a very short time. When Zn is used as a reducing agent, the reduction reaction can be completed in one minute, which is much more efficient than the traditional iron powder. In

2.4 Synthesis of Graphene Oxide—Reduced Graphene Oxide

addition, the GO obtained by using Zn as reductant not only reacted quickly but also had high quality, and the C/O ratio is 33.5, which indicated that it restored the graphene-like structure [28].

On the other hand, when Al is used as a reducing agent, it usually takes only 30 min to complete the reduction reaction. This is due to the standard potential of Al being -1.68 v ($Al = Al^{3+} + 3e^{-}$) [28]. The conductivity of rGO obtained by reduction is 2.1×10^3 S m^{-1}, only one order of magnitude lower than that of the original graphite (approximately 3.2×10^4 S m^{-1}). This indicates the high efficiency and effectiveness of Al and Zn as reducing agents [29].

Solvent

Organic solvents play a crucial role in the reduction of GO, including various solvents such as N,N-dimethylformamide (DMF), tetrahydrofuran (THF), isopropanol, benzyl alcohol, and methanol. Among these solvents, benzyl alcohol stands out for its excellent performance. Research indicates that using benzyl alcohol as a reducing agent can achieve the highest electrical conductivity of rGO, while significantly enhancing the stability and reduction degree of the colloidal solution. Selecting the appropriate solvent during the reduction process is essential for achieving optimal performance in rGO [30].

The removal of oxygen-containing functional groups in GO by DMF is a novel and efficient method, and rGO with excellent electrical properties can be prepared in 30 min [30]. By dispersing GO in DMF and then heating it to the boiling point of DMF (153 °C), local pressure and temperature rise are generated in the reactor vessel. DMF is not only an effective GO solvent, but also a precursor of carbon monoxide (CO). CO as a strong reducing agent contributes to the reduction process of GO. The prepared rGO exhibits a very high conductivity of about 6380 S m^{-1} [31]. These advances highlight the potential of organic solvents in the optimization of GO reduction processes and provide new ways to generate materials suitable for a variety of applications.

Ascorbic acid and NaBH$_4$

Ascorbic acid, as an effective reducing agent for GO, is widely used because of its environmental and economic characteristics. 50 ml GO solution is mixed with 0.1 m ascorbic acid, heated to 70 °C and stirred for two hours, followed by centrifugation to obtain rGO. The absorption peak shifted from 226 to 260 nm, indicating that the oxidation functional group is removed, demonstrating the high efficiency of ascorbic acid as a reducing agent [32]. 25 mg GO is treated with sodium borohydride as a reducing agent, dissolved in 50 ml deionized water, sonicated and pH adjusted to 10–11. Then, add 200 mg of sodium borohydride and stir the mixture at 95 °C for 1 h. It is then filtered, washed, and dried. The C/O ratio of rGO prepared by this method is 6.69, which is lower than that of rGO prepared by ascorbic acid (7.62), indicating that the reduction effect is relatively weak [33].

2.4.2 Biological Pathways

Plant extracts

Green synthesis technology uses environmentally friendly and sustainable methods and materials, significantly reducing the impact on the environment. This technique involves using extracts from different parts of the plant as key components of the synthesis process. Commonly used plant extracts include tea leaves, green tea, citrus peels, sugar cane, and palm oil leaves [34, 35]. Palm oil leaf extract has been utilized to reduce GO, offering an economical and environmentally friendly approach. This research effectively removed the hydroxyl groups in graphene oxide, thereby increasing the ratio of carbon to oxygen (from 1:1 to 3:1). The successful reduction of GO is confirmed by X-ray diffraction and Raman spectroscopic analysis. This discovery not only offers an environmentally friendly method for the production of high-value materials utilizing agricultural waste, but also highlights the primary advantage of using plant extracts to rGO: the resulting rGO exhibits exceptional colloidal stability and high biocompatibility in both organic and polar solvents [36].

Polysaccharides and natural proteins

Polysaccharides and proteins such as D-fructose, glucan, glycine, sucrose, bovine serum albumin (BSA), L-glutathione (GSH), dopamine, melatonin are explored as a natural reducing agent [37]. GSH has been used as a reducing agent to convert GO to rGO. The interaction of GSH with GO triggers oxidation, ultimately leading to the formation of dimeric glutathione disulfide (GSSG). This compound is characterized by two GSH molecules connected through a disulfide bond, accompanied by the release of protons (Fig. 2.3). These released protons react with oxygen-containing functional groups on GO, converting it into rGO in a mild aqueous environment. The process requires mixing GSH and GO in solution, followed by sonication and heating, ultimately producing stable, dispersed graphene nanosheets [38].

Microorganism

The use of pathogens and microorganisms such as yeast, especially Shewanella, as reducing agents to synthesize rGO has attracted widespread attention. The metal reduction ability of Shewanella is particularly effective under various conditions, focusing on reducing C–OH groups instead of C=O groups, due to the higher bond energy of C=O. This process is mainly attributed to the main protein component of its Mtr respiratory pathway [1]. The comparative study of Shewanella MR-1 highlighted the different roles of specific proteins in GO reduction. It is found that the use of riboflavin and anthraquinone-2,6-disulfonic acid (AQDS), both with electron shuttling ability, can significantly improve the reduction efficiency of the MR-1 strain [39].

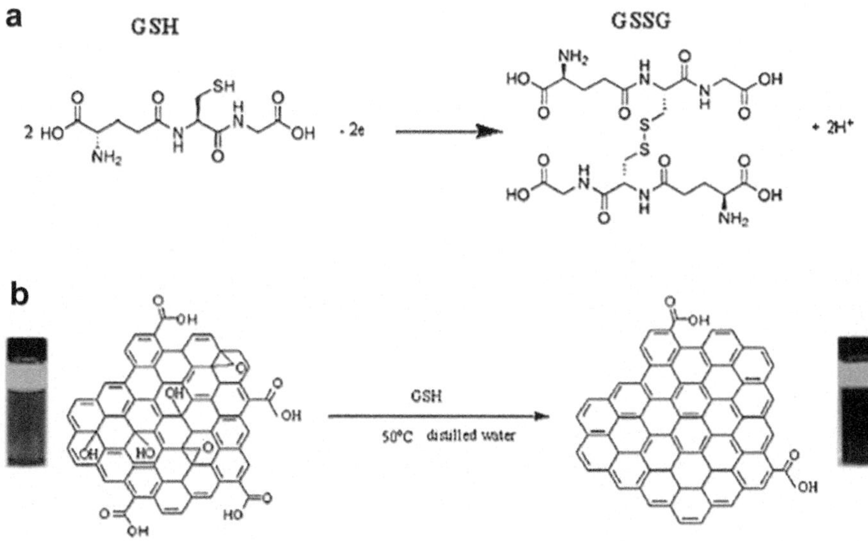

Fig. 2.3 Route to synthesize graphene from GO: **a** each GSH releases a proton and reacts with another glutathione to form GSSG and **b** GO is reduced to form graphene in one step, reproduced with permission from Ref. [38]

2.4.3 Photoreduction

Photoreduction technology uses light energy (laser, camera flash, plasma, or ultraviolet light) to remove oxygen-containing functional groups in GO and achieve its reduction. The key to this approach is that light energy is converted into heat, which in turn induces thermal reduction, making the photoexcited and heat-mediated reduction processes essentially the same [40, 41]. Photoreduction offers several advantages over chemical or thermal reduction, including room temperature processing, mask-free patterning, non-contact processing, no need for chemical reagents, and a controllable reduction process [40, 42].

Laser

Currently, one of the most advanced and widely used technologies, is the use of femtosecond lasers for photoreduction of GO, especially applications based on direct writing and beam interference technologies [1]. Femtosecond laser processing is regarded as an ideal choice for industrial development due to its simplicity, mask-free, and high efficiency. Femtosecond laser plasma lithography (FPL) technology has been used to successfully achieve high-speed, large-area micro/nanoscale manufacturing and photoreduction (300 nm SiO_2/Si) of GO thin films (~ 140 nm) on silicon substrates. Remarkably, even on the centimeter scale, the surface of this reduced GO film can form a periodic grating structure with a high degree of regularity (~ 680 nm), demonstrating the application of femtosecond laser technology in the fields of precision manufacturing and photoreduction, which shows huge potential for future industrial-scale applications [41].

Plasma

Plasma treatment, as an innovative and environmentally friendly surface treatment method, is used to optimize the microstructure, adhesion performance and electrical properties of GO films. A constant 300 W radio frequency (RF) power is utilized for oxygen plasma treatment, with treatment times ranging from 0 to 7 min. The study showed that the 5-min treatment significantly reduced the proportion of oxygen-containing groups (epoxy groups, ketone groups, and carboxyl groups) in the GO film, from 48.8 to 33.56%. The surface roughness (Ra) increased from about 7.8 μm to about 8.7 μm, and the adhesion force enhanced to 134.84 mN m^{-1}. After 7 min of treatment, thermogravimetric analysis (TGA) showed that the weight loss reached 51.66%. The addition of C–O bonds during processing also increased the conductivity to 0.156 S m^{-1}. These results show that RF oxygen plasma treatment can effectively reduce the oxygen-containing groups in GO film, thus improving its physical and electrical properties and making it better meet the needs of various applications [40].

Photocatalytic

UV-activated photoinitiators and photoactivated inorganic nanoparticles are two main methods for the photocatalytic reduction of GO. The key to the reduction process is the generation of free radicals, electrons, or electron–hole pairs, which promote the reduction of GO. Both of these technologies efficiently produce the required active substances and promote GO reduction [1, 8]. GO is functionalized through thioether free-radical addition (TERA) using a photoinitiator (PI). The results show that the degree of functionalization of GO increases with increasing PI concentration, but the optimal PI concentration is 1.4 mM (Fig. 2.4) [42].

2.4.4 Thermal Reduction

Thermal reduction is an effective technology for reducing GO, especially in large-scale preparation. This method requires heating GO in an inert gas environment to convert its carbon atoms from the oxidation state to the reduction state, thus forming a solid material. In the reduction process, the oxygen functional groups of GO are gradually removed, mainly forming carbon oxides such as CO_2 and CO as by-products. This thermal reduction technique has a profound effect on the structure and properties of materials [43].

Hydrothermal

The hydrothermal reduction of GO involves treating it under high temperature and pressure conditions to convert it into rGO. RGO is prepared on a polyvinylidene fluoride (PVDF) substrate using vacuum filtration technology (Fig. 2.5) [44]. With the extension of hydrothermal treatment time, the sp^2 carbon domain in GO is gradually

2.4 Synthesis of Graphene Oxide—Reduced Graphene Oxide

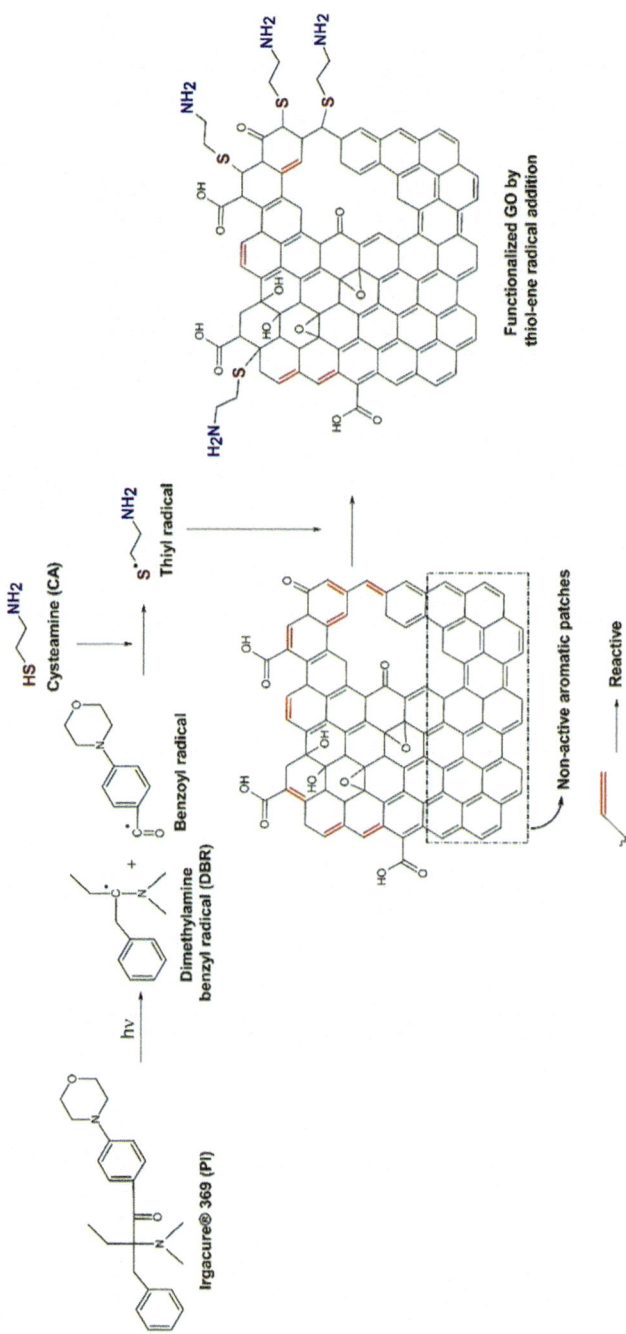

Fig. 2.4 Irgacure® 369 undergoes photocleavage to produce radicals that can functionalize GO through TERA by either abstracting hydrogen from CA or adding to the alkene in GO, reproduced with permission from Ref. [42]

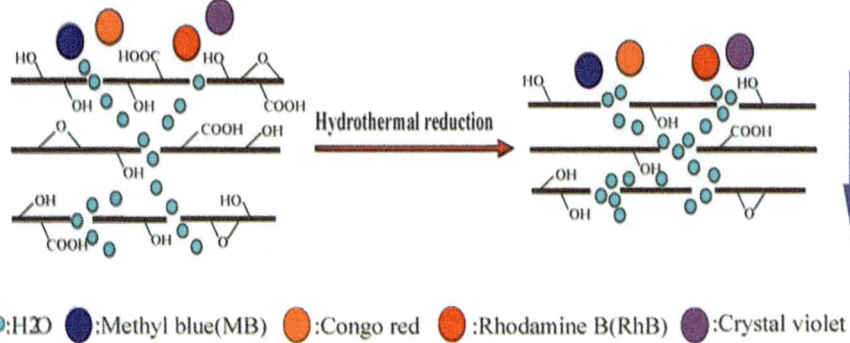

●:H2O ●:Methyl blue(MB) ●:Congo red ●:Rhodamine B(RhB) ●:Crystal violet

Fig. 2.5 Process of hydrothermal reduction of GO, reproduced with permission from Ref. [44]

restored, and at the same time, the oxygen-containing functional groups (hydroxyl, carboxyl, and epoxy groups) gradually disappear, thereby achieving effective reduction of rGO. The research results showed that the rGO membrane prepared after 9 h of hydrothermal treatment showed excellent performance in terms of water permeability and rejection rate, especially for dyes such as methyl blue, Congo red, and crystal violet. This work provides a simple and green strategy for preparing rGO membranes with separation functionality [44].

Microwave-assisted reduction

Microwave (MW)-assisted reduction of GO is a method that utilizes microwave radiation and can achieve efficient reduction of GO in a relatively short time. The MW reduction strategy mainly includes three steps [1]:

(i) MW-mediated thermal reduction of GO.
(ii) MW-mediated chemical reduction of GO.
(iii) MW-mediated simultaneous stripping and reduction of GO.

A microwave-assisted method is used for the rapid synthesis of rGO using sodium citrate (NaC) as the reducing agent. It is observed that at a synthesis temperature of 140 °C, the rGO sample exhibits the highest carbon surface area (C sp) and its specific capacitance reaches 182.5 F g^{-1} (293%) at a current density of 0.1 A g^{-1} [45]. This sample has excellent energy storage properties compared with other rGO samples reduced at higher temperatures. It is worth noting that oxygen-containing functional groups still exist on the rGO product synthesized at low temperature, which helps to promote ionic interactions with the electrolyte, thus enhancing its charge storage capability.

2.5 Summary

To explore various strategies for reducing GO to rGO, including chemical, biological, thermal, and photoreduction methods, highlighting their distinct properties and reduction mechanisms, it systematically outlines:

(i) A systematic analysis of the characteristics and mechanisms of different reduction methods has deepened the understanding of the rGO synthesis process.
(ii) The importance of selecting an appropriate reduction technique in the rGO synthesis process has been emphasized.
(iii) This study makes significant contributions to the fields of materials science and nanotechnology, advancing progress toward more environmentally friendly and efficient rGO synthesis technologies.

References

1. Agarwal V, Zetterlund PB (2021) Strategies for reduction of graphene oxide–a comprehensive review. Chem Eng J 405:127018
2. Abdallah IA et al (2023) Applications of layered double hydroxides in sample preparation: a review. 192:108916
3. Zhou Q et al (2024) Recent progress in magnetic polydopamine composites for pollutant removal in wastewater treatment. 130023
4. Tamang S et al (2023) A concise review on GO, rGO and metal oxide/rGO composites: fabrication and their supercapacitor and catalytic applications. 947:169588
5. Xiao Y et al (2023) Synthesis and functionalization of graphene materials for biomedical applications: recent advances, challenges, and perspectives. 10(9):2205292
6. Hulagabali MM, Vesmawala GR, Patil YD (2023) Synthesis, characterization, and application of graphene oxide and reduced graphene oxide and its influence on rheology, microstructure, and mechanical strength of cement paste. 71:106586
7. Sun Y-B et al (2023) One-step synthesis of S, N dual-element doped rGO as an efficient electrocatalyst for ORR. 940:117489
8. Ahmed A et al (2023) Synthesis techniques and advances in sensing applications of reduced graphene oxide (rGO) Composites: a review. Compos A Appl Sci Manuf 165:107373
9. de Barros NG et al (2023) Graphene oxide: a comparison of reduction methods. 9(3):73
10. Deshwal N et al (2023) A review on recent advancements on removal of harmful metal/metal ions using graphene oxide: experimental and theoretical approaches. 858:159672
11. Gao B et al (2024) Graphene-based aerogels in water and air treatment: a review. 149604
12. Nazari N et al (2023) The effect of phosphorus and nitrogen dopants on structural, microstructural, and electrochemical characteristics of 3D reduced graphene oxide as an efficient supercapacitor electrode material. 137:110144
13. Boddula R et al (2024) Fabricating and designing graphene-based nanomaterials using different current 'top-down' and 'bottom-up' techniques. Graphene-based nanomaterials. CRC Press, pp 33–46
14. Shi L, Wang B, Lu SJM (2023) Efficient bottom-up synthesis of graphene quantum dots at an atomically precise level. 6(3):728–760
15. Wen Y et al (2023) Preparation of graphene by exfoliation and its application in lithium-ion batteries. 961:170885

16. Borane N et al (2024) Recent trends in the "bottom-up" and "top down" techniques in the synthesis and fabrication of myriad carbonaceous nanomaterials. Carbon-based nanomaterials in biosystems. Elsevier, pp 91–120
17. Pham PV et al (2024) Layer-by-layer thinning of two-dimensional materials
18. Alam SN, Sharma N, Kumar L (2017) Synthesis of graphene oxide (GO) by modified hummers method and its thermal reduction to obtain reduced graphene oxide (rGO). Graphene 6(1):1–18
19. Stankovich S et al (2007) Synthesis of graphene-based nanosheets via chemical reduction of exfoliated graphite oxide. Carbon 45(7):1558–1565
20. Qiang Y et al (2024) A new suggestion to marine gold extraction: utilizing reduced graphene oxide membranes within seawater desalination processes. 174602
21. Xi G et al (2022) Insights into the degradation of carbamazepine by persulfate activated by chalcopyrite: degradation mechanism and synergy with zero-valent iron
22. Chasanah U et al (2022) Role of temperature and time exposure for controlled and accelerated synthesis of graphene oxide using tour method. 22(5):1205–1217
23. Wambu EW, Huang J (2024) Chemistry of graphene: synthesis, reactivity, applications and toxicities. BoD–Books on Demand
24. Agarwal V, Zetterlund PB (2021) Strategies for reduction of graphene oxide—a comprehensive review. 405:127018
25. De Silva KKH et al (2020) Restoration of the graphitic structure by defect repair during the thermal reduction of graphene oxide. 166:74–90
26. Gul W et al (2023) Synthesis of graphene oxide (GO) and reduced graphene oxide (rGO) and their application as nano-fillers to improve the physical and mechanical properties of medium density fiberboard. 10:1206918
27. Ren P-G et al (2010) Temperature dependence of graphene oxide reduced by hydrazine hydrate. Nanotechnology 22(5):055705
28. Mei X, Ouyang J (2011) Ultrasonication-assisted ultrafast reduction of graphene oxide by zinc powder at room temperature. Carbon 49(15):5389–5397
29. Fan Z et al (2010) An environmentally friendly and efficient route for the reduction of graphene oxide by aluminum powder. Carbon 48(5):1686–1689
30. Tarcan R et al (2020) A new, fast and facile synthesis method for reduced graphene oxide in N,N-dimethylformamide. Synth Met 269:116576
31. Ai K et al (2011) A novel strategy for making soluble reduced graphene oxide sheets cheaply by adopting an endogenous reducing agent. J Mater Chem 21(10):3365–3370
32. Hessain HA, Hassan J (2020) Green synthesis of reduced graphene oxide using ascorbic acid. Iraqi J Sci 1313–1319
33. Bansal K, Singh J, Dhaliwal A (2022) Synthesis and characterization of graphene oxide and its reduction with different reducing agents. In: IOP conference series: materials science and engineering. IOP Publishing
34. Shafey AME (2020) Green synthesis of metal and metal oxide nanoparticles from plant leaf extracts and their applications: a review. Green Process Synth 9(1):304–339
35. Soni V et al (2021) Sustainable and green trends in using plant extracts for the synthesis of biogenic metal nanoparticles toward environmental and pharmaceutical advances: a review. Environ Res 202:111622
36. Faiz MA et al (2020) Low cost and green approach in the reduction of graphene oxide (GO) using palm oil leaves extract for potential in industrial applications. Results Phys 16:102954
37. Norahan MH et al (2023) Structural and biological engineering of 3D hydrogels for wound healing. Bioact Mater 24:197–235
38. Pham TA et al (2011) One-step reduction of graphene oxide with L-glutathione. Colloids Surf, A 384(1–3):543–548
39. Jiao Y et al (2011) Deciphering the electron transport pathway for graphene oxide reduction by Shewanella oneidensis MR-1. J Bacteriol 193(14):3662–3665
40. El-Hossary F et al (2021) The effective reduction of graphene oxide films using RF oxygen plasma treatment. Vacuum 188:110158

41. Zou T et al (2020) High-speed femtosecond laser plasmonic lithography and reduction of graphene oxide for anisotropic photoresponse. Light Sci Appl 9(1):69
42. Piñeiro-García A et al (2022) Functionalization and soft photoreduction of graphene oxide triggered by the photoinitiator during thiol-ene radical addition. FlatChem 33:100349
43. Silva Filho J et al (2020) A thermal method for obtention of 2 to 3 reduced graphene oxide layers from graphene oxide. SN Appl Sci 2:1–8
44. Fan X et al (2020) Hydrothermal reduced graphene oxide membranes for dyes removing. Sep Purif Technol 241:116730
45. Tamang S et al (2023) Microwave-assisted reduction of graphene oxide using Artemisia vulgaris extract for supercapacitor application. J Mater Sci: Mater Electron 34(7):575

Chapter 3
Reduced Graphene Oxide Thin Film Electrode

Abstract Reduced graphene oxide (rGO) electrodes have garnered significant attention in the fields of energy storage, sensors, and flexible electronics due to their high electrical conductivity, mechanical strength, and excellent flexibility. This chapter evaluates various rGO film preparation technologies, such as spin coating and vacuum filtration, with a focus on addressing challenges related to quality and large-scale production. The specific effects of different preparation techniques on the properties of rGO films are discussed. This chapter provides valuable insights into the preparation and applications of rGO films, offering a foundation for future research and development in these areas.

Keywords Reduced graphene oxide · Fabrication techniques · Thin films · Quality control · Scalable production · Application areas

3.1 Introduction

In the research field of nanomaterials and advanced composite materials, reduced graphene oxide (rGO) films have become a research hotspot due to their excellent electronic and mechanical properties [1], exhibiting extremely high electrical conductivity, with conductivity reaching up to 87,100 S m^{-1}, a sheet resistance of 21.2 Ω s^{-1} q^{-1}, and a charge mobility of 16.7 cm^2 V^{-1} s^{-1} [2]. In addition, rGO films have a high specific surface area (2630 m^2 g^{-1}) and excellent conductivity (from thousands to tens of thousands of S m^{-1}); it has a broad application prospect in the fields of energy storage, flexible electronic devices, and sensor technology [3, 4]. However, there are still many challenges in the preparation process, including the diversity of preparation methods, the feasibility of large-scale production, and the difficulty of high-quality control [3].

In this chapter, various preparation techniques of rGO films are analyzed and discussed, including spin coating, vacuum filtration, spray coating, dip coating, drop coating, and electrophoretic deposition. The validity of these methods in laboratory

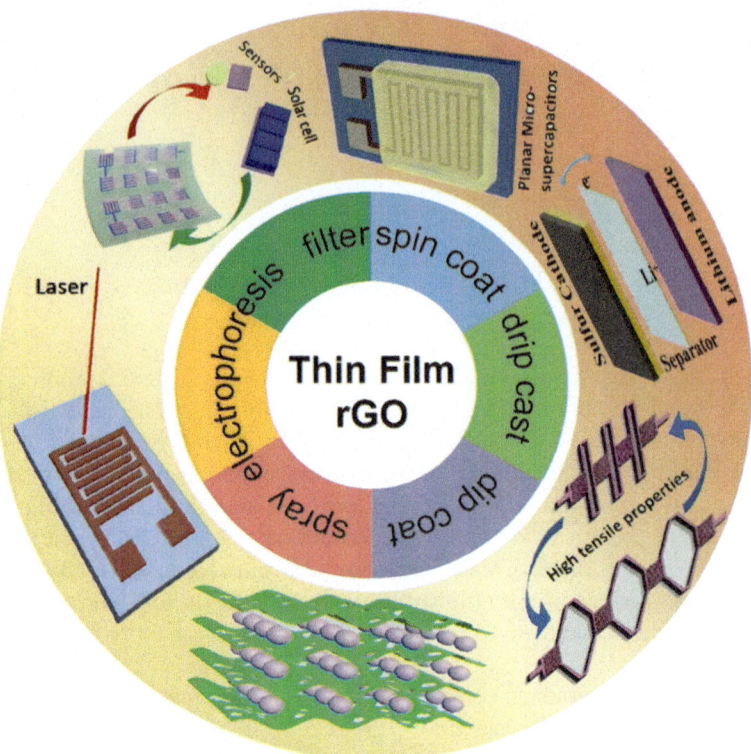

Fig. 3.1 Synthesis and application of rGO films

scale is further discussed, and their feasibility in industrial production is fully considered, with a view to better application in the fields of energy storage, sensors, and flexible electronic materials (Fig. 3.1).

3.2 Preparation Method of Reduced Graphene Oxide Film

In the field of nanomaterials, rGO films have become a hot research topic due to their excellent physical and chemical properties, such as high conductivity, high mechanical strength, and excellent flexibility [5]. The research shows that rGO film shows great potential in advanced applications such as energy storage, electronic devices and sensors, and can be comparable with other top materials. RGO films have many preparation technologies, each of which have its unique advantages and limitations, and are suitable for different applications. In this chapter, several common rGO membrane preparation techniques are introduced, including spin coating, vacuum filtration, spray coating, dip coating, drop casting, and electrophoresis [6–8].

3.2 Preparation Method of Reduced Graphene Oxide Film

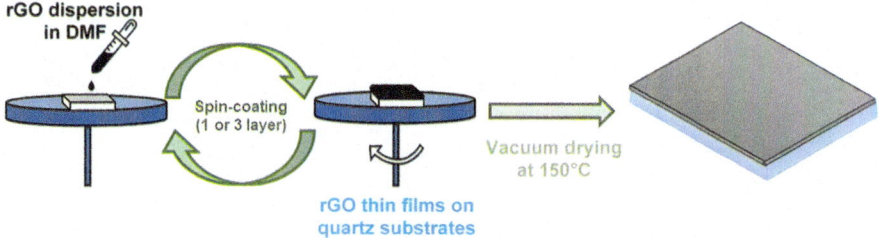

Fig. 3.2 Process of preparing rGO films by spin-coating method, reproduced with permission from Ref. [11]

3.2.1 Spin Coating

Spin coating is a process that relies on the centrifugal force and gravity generated by the rotation of the workpiece to distribute the coating droplets uniformly across the surface of the workpiece [9]. A quartz substrate (2.5 cm × 2.5 cm × 2.5 mm) is selected as the substrate, and the substrate is first properly cleaned in acetone and distilled water, followed by ultrasonic treatment to ensure deep cleaning [10]. The cleaned substrate is immersed in Piranha solution with a ratio of 3:1 to enhance the adhesion of the substrate surface. Then, rGO is coated on the quartz substrate and rotated using a spin coater at 600 revolutions per minute (RPM) for 30 s to achieve a uniform coating effect. Afterward, the rGO-coated film is placed in a vacuum environment and dried at 150 °C for 3 h to remove the remaining solvent (Fig. 3.2) [11]. The advantage of the spin-coating method is that it is simple to operate and easy to control the thickness of the film, but there are certain limitations in the preparation of large-area films. The rGO films synthesized by this method are mainly used for small batch production and development and are suitable for laboratory-scale electronic device manufacturing.

3.2.2 Vacuum Filtration

Vacuum filtration is a technique that utilizes suction and gravity to pass liquid through a filter [12]. First, the rGO powder is dispersed in water or organic solvent, and ultrasonic treatment technology is used to ensure that the rGO powder is evenly dispersed in the solvent. Then, a commercial membrane such as polyvinylidene fluoride (PVDF), nylon, or polycarbonate (PC) is used as the filter [13]. The filter is placed at the bottom of a funnel or similar device connected to a vacuum pump. The rGO dispersion is vacuum filtered at a pressure difference of 730 mm Hg. During this process, rGO sheets accumulate on the surface of the filter membrane to form an rGO membrane. Finally, after filtration is complete, turn off the vacuum pump, transfer the film to the target substrate (quartz, silicon wafer, or other materials), and dry it at room temperature overnight to remove residual solvent [6, 13]. Vacuum filtration can produce large-area and uniform films, but the preparation process is relatively complex and may cause damage or defects during film transfer. The rGO films produced by this method are mainly used in the fields of large-area sensors, membrane separation technology and energy storage [6].

3.2.3 Electrostatic Spraying

Electrostatic spraying (ESD) is a coating method that uses a high-voltage electrostatic field to direct negatively charged paint particles toward the workpiece surface, where they are attracted and adhere to it [14]. In this process, rGO suspensions are prepared into droplets suitable for dispersion. The size and charge of the nozzle are controlled by the flow and voltage of the precursor between the nozzle and the substrate. Using electrostatic forces, these charged droplets are accurately guided and sprayed onto the substrate, and their temperature has a significant effect on film properties (Fig. 3.3) [7, 15]. Unlike other methods, rGO films prepared by ESD technology do not require heat treatment after fabrication. This technique shows great potential in the preparation of rGO films with special functions and is suitable for various advanced applications such as large solar panels, large-area flexible electronic devices, and some special protective coatings [7].

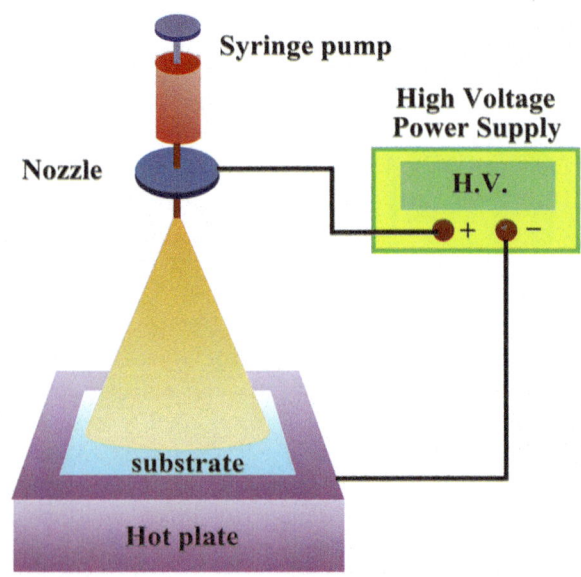

Fig. 3.3 Schematic diagram of electrostatic spray deposition system, reproduced with permission from Ref. [7]

3.2 Preparation Method of Reduced Graphene Oxide Film

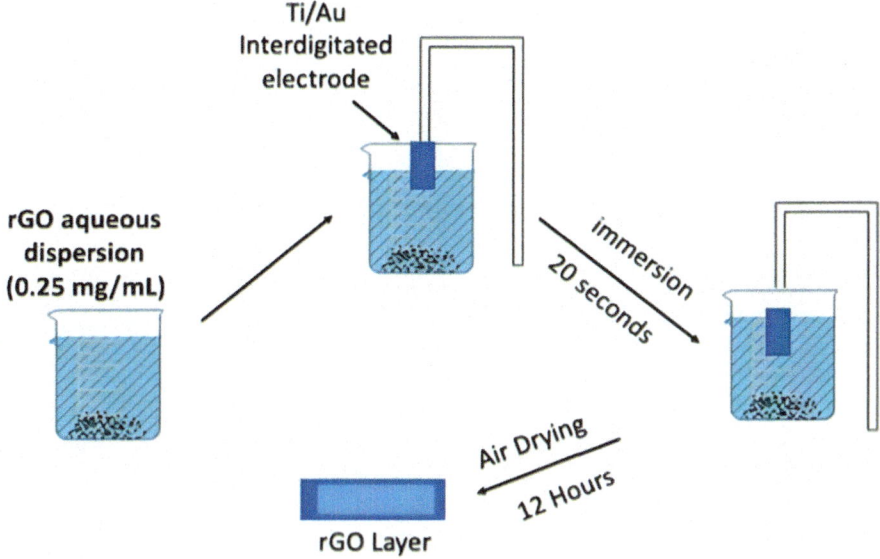

Fig. 3.4 Schematic diagram of preparation of rGO film by dip-coating method, reproduced with permission from Ref. [17]

3.2.4 Dip Coating

Dip coating is a sophisticated method to prepare rGO films with excellent sensing properties [16]. First, rGO powder is uniformly dispersed in water or other organic solvents at a concentration of 0.25 mg mL^{-1} and is treated with sonication to enhance its dispersion in the solvent. Next, silicon/silicon dioxide (Si/SiO$_2$) with good electrical conductivity and stability is selected as the substrate, and clean titanium/gold (Ti/Au) interconnected electrodes are used. Finally, the electrode is immersed vertically in a beaker containing the rGO aqueous dispersion. After immersing for 20 s, the electrode is taken out of the solution, and then heat treated and dried overnight at room temperature and pressure to remove residual solution on the surface (Fig. 3.4) [16, 17]. This method is not only simple to operate, but also easy to control the thickness of the film, and can prepare uniform, continuous rGO films with good sensing performance. This rGO film is suitable for many fields such as biomedical coatings, anti-corrosion coatings, and sensors [17].

3.2.5 Drop Casting

Drop casting involves depositing a solution onto a substrate and then allowing the solution to evaporate, leaving behind a thin film [18]. Silicon nanowire (SiNW)

Fig. 3.5 Process of preparing rGO films by drop casting method, reproduced with permission from Ref. [2]

arrays are first etched by applying chemical etching technology to the silicon wafer. Subsequently, GO powder and ethanol solution are mixed and sonicated to form a suspension (concentration of 0.1–1 mg/ml), which was uniformly drop-cast on top of the silicon nanowire array at approximately 75 μl per square centimeter. After the suspension is fully spread and covers the top surface of the nanowire array, the resulting nanowire array is dried at 60 °C for three minutes, and then the next round of drop casting is performed. After repeated drop casting and drying for many times, it was dried at room temperature overnight and finally annealed at 1000 °C for 3 h in a mixed atmosphere of Ar (95%)–H_2 (5%) to successfully prepare an rGO film (Fig. 3.5) [2]. The rGO film produced by this method exhibits excellent electrical properties, with a conductivity of up to 87,100 S/m, a sheet resistance of only 21.2 Ω s^{-1} q^{-1}, and a mobility of 16.7 cm^2 V^{-1} s^{-1} [2]. The drop casting method is simple to operate and suitable for rapid prototyping and small batch production. It is especially suitable for high-performance electronic devices such as laboratory-grade sensors, prototype batteries, and supercapacitors [8].

3.2.6 Electrophoresis

Electrophoretic deposition is a process in which an electric field is applied to deposit charged particles from an electrolyte solution onto the surface of an electrode with the opposite charge [19, 20]. First, GO powder is fused with an aqueous solution and sonicated to form a suspension (concentration: 0.25 mg ml^{-1}). Direct current (DC) magnetron sputtering technology is then used to accurately deposit a titanium film on the quartz substrate. In the electrophoretic deposition device, an anode composed of platinum sheets and a cathode composed of titanium sheets are immersed in a GO suspension. These components are positioned parallel to each other with a separation of 1 cm to facilitate electrophoretic deposition. The material is electrophoresed for 1 min using a constant voltage of 10 V while measuring the current varying between 0.1 and 1.0 mA. After completion of deposition, the resulting GO film is heated at 400 °C for 20 min in an argon atmosphere to achieve drying and annealing and successfully prepare an efficient rGO film (Fig. 3.6) [21, 22]. The rGO produced by this method shows great potential in the field of optoelectronics due to its excellent

3.3 Application of Reduced Graphene Oxide Film

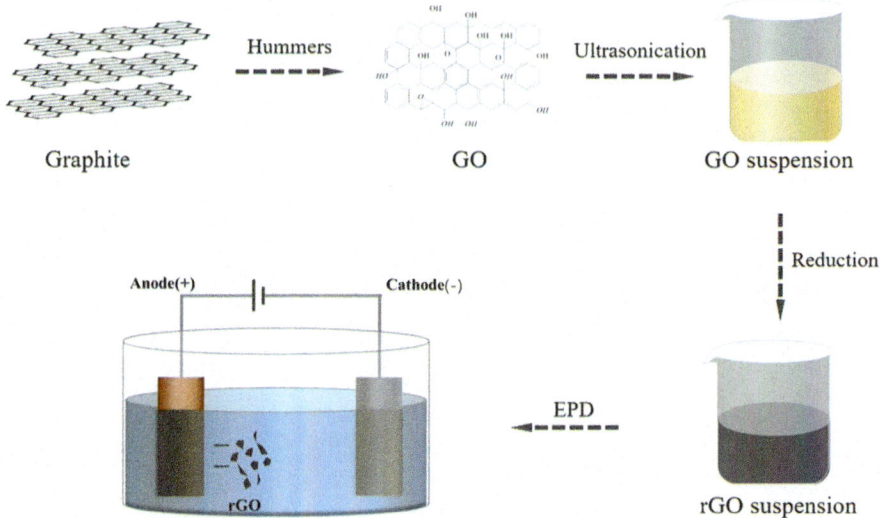

Fig. 3.6 Process of preparing rGO thin film by electrophoretic deposition method, reproduced with permission from Ref. [21]

refractive index, optical quality, and transparency. It is mainly suitable for high-precision electronic devices, nanoscale composite materials, and catalyst carriers with specific microstructure requirements [21].

3.3 Application of Reduced Graphene Oxide Film

3.3.1 Energy Storage

With its unique physical and chemical properties, especially its high specific surface area (2630 m^2 g^{-1}) and excellent electrical conductivity (thousands to tens of thousands S m^{-1}), rGO films have become a promising candidate in the field of energy storage [3, 4]. These materials exhibit broad application prospects in energy storage fields such as batteries and supercapacitors (Fig. 3.7) [23].

Research shows that rGO films prepared by reducing GO using caffeic acid (CA) exhibit excellent electrochemical properties in supercapacitors. Even at high scan rates, significant specific capacitances are achieved (96 F g^{-1} in aqueous solution and 74 F g^{-1} in organic electrolytes) [24].

The α-Fe_2O_3/rGO composite material synthesized using in situ synthesis and mechanical stirring methods can be used as an electrode material for supercapacitors. The composite material exhibits excellent electrochemical properties. At a current

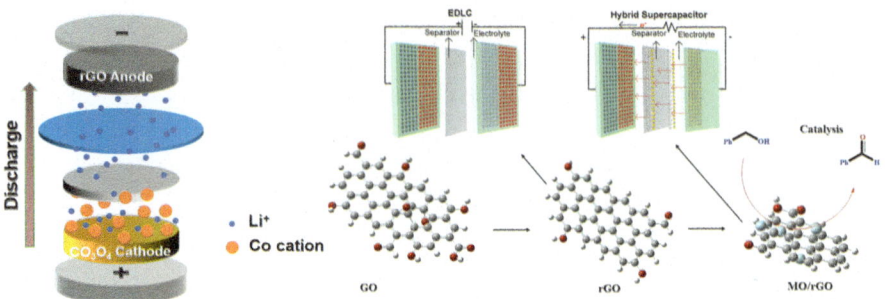

Fig. 3.7 RGO used as an anode in lithium-ion batteries and applications of rGO in supercapacitors, reproduced with permission from Ref. [23]

density of 1 A g^{-1}, the specific capacitance is 970 F g^{-1}. Even after 2000 cycles at a current density of 5 A g^{-1}, 75% of the capacitance was retained [25].

A unique symmetric supercapacitor (SSC) device architecture is developed by using rGO and nitrogen-doped rGO (N-rGO) electrodes. This SSC exhibits a wide voltage window (2.2 V), high energy density (106.3 W h kg^{-1}), and extremely high-power density (15,184.8 W kg^{-1}). It has high stability, and the capacity can still maintain 95.5% after 10,000 charge and discharge tests, and the retention rate is 90.5% after an 8-h voltage test [26].

3.3.2 Sensor

The rGO films also have broad application prospects in the sensor field. Its high specific surface area and abundant functional groups make it an ideal sensing material capable of efficiently capturing and identifying various chemical substances [27].

An innovative rGO-based gas sensor enhanced with peptide receptors enables accurate detection of trinitrotoluene 2,4-dinitrotoluene (DNT), a major by-product of trinitrotoluene (TNT). This advanced sensor exploits differential signals of DNT-specific and non-specific binding peptides, a technology that ensures superior selectivity and advanced perception. This high specificity is further verified by its unique response to a range of substances such as DNT, acetone, toluene, and ethanol, making it a breakthrough tool in the field of explosives detection (Fig. 3.8) [28].

It is a simple, economical, and efficient method to develop an independent rGO film optical sensor. The sensor developed using it shows the highest sensitivity at various temperatures for irradiation at a wavelength of 635 nm. It is worth noting that at 123 K, the sensitivity of a 635 nm laser with a power density of 1.4 MW/mm^2 is 49.2%. In addition, the study also shows that the sensor can effectively move from visible light to near-infrared region, and its performance is especially remarkable at low temperature [29].

3.3 Application of Reduced Graphene Oxide Film

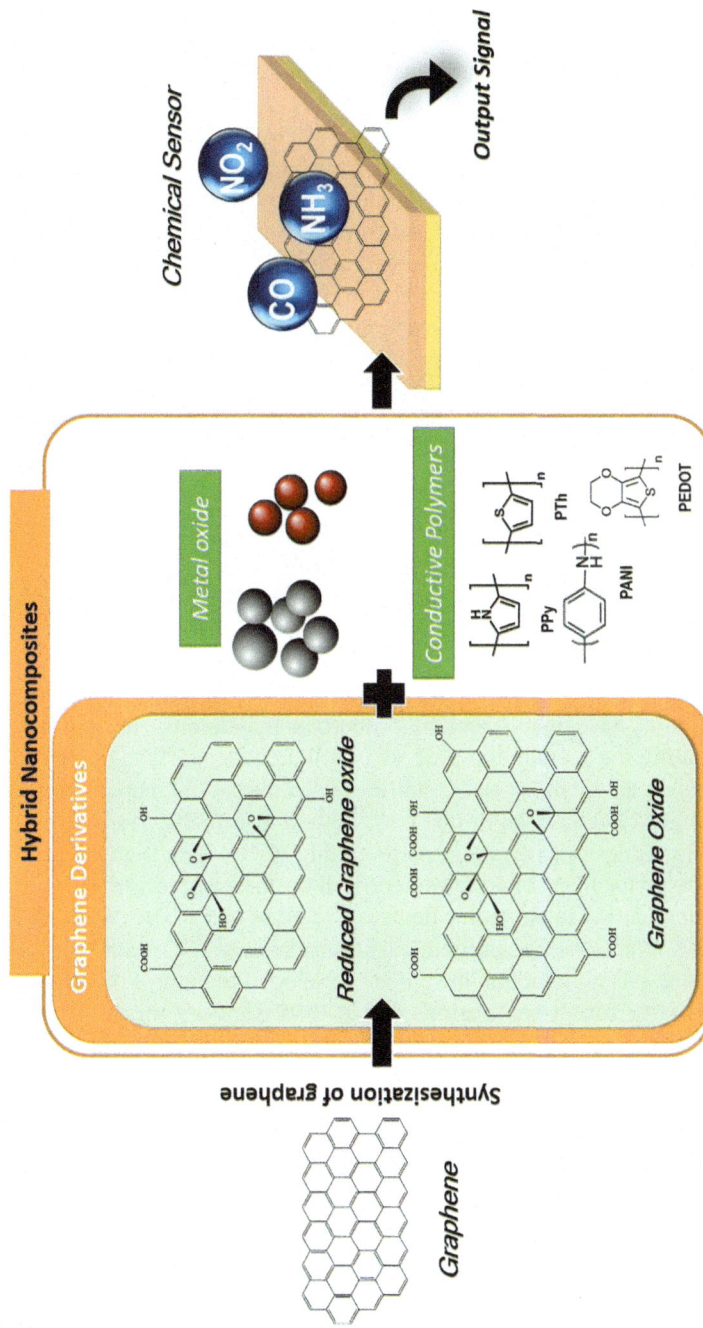

Fig. 3.8 Application of rGO-based composites in sensors, reproduced with permission from Ref. [28]

3.3.3 Flexible Electronics and Smart Materials

With the development of science and technology, rGO films have been widely used in flexible electronics and smart materials because of their excellent electrical, mechanical, optical, and thermal properties [30, 31].

Recent research has demonstrated the use of rGO films as anode materials, and the successful development of an innovative supercapacitor based on bamboo fiber by combining MnO_2-$NiCo_2O_4$ (cathode material) and LiCl/PVA gel (solid electrolyte). This textile-based MnO_2-$NiCo_2O_4$//rGO asymmetric supercapacitor not only exhibits excellent electrochemical performance but also has high capacitance (2.12 F/cm^2), excellent energy density (37.8 mW/cm^3), excellent maximum power, density (2678.4 mW/cm^3), and excellent cycling stability. The most outstanding thing is that the supercapacitor still maintains its electrochemical properties despite undergoing various mechanical deformations, which fully proves its excellent flexibility and high mechanical strength, and indicates that rGO films have great application potential in the field of flexible electronic devices [30].

3.4 Current Challenges and Future Directions

3.4.1 Quality Control

Although rGO films have many excellent properties, maintaining the consistency of their quality during the preparation process is still a challenge. The quality of rGO largely depends on the methods and conditions used during the reduction process.

The type of glass wafer used in one new method affects the quality of the rGO produced, with magnesium silicate glass providing rGO with a high C/O ratio and electrical conductivity [32]. In addition, controlling the oxygen content in rGO is crucial for its application in electronic devices and other fields. The degree of reduction significantly affects the electrochemical performance of rGO. Although methods such as hydrazine-based reduction have been developed to control oxygen content, they are not suitable for all applications due to the toxicity of the reducing agents used [33].

3.4.2 Large-Scale Production

Although the laboratory-scale preparation of rGO films is relatively mature, scaling it up to commercial-scale production still faces challenges. Achieving homogeneity of rGO films while scaling up production is a significant challenge. Producing large and uniform rGO films is crucial for its applications in various fields. New methods

3.4 Current Challenges and Future Directions

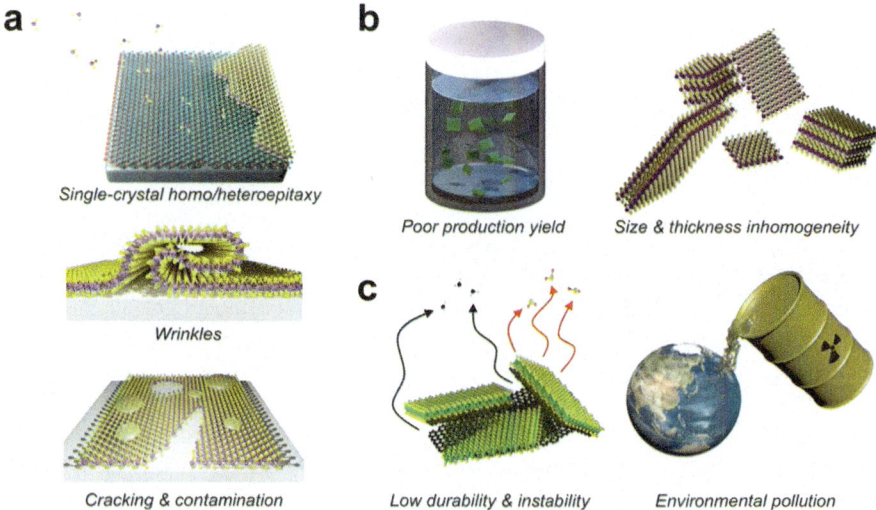

Fig. 3.9 Current challenges in the mass production of 2D materials, reproduced with permission from Ref. [34]

such as self-assembly processes have been developed to address this challenge, but they have their own limitations [33].

Three mainstream large-scale production technologies include chemical vapor deposition (CVD), liquid phase exfoliation and wet chemical synthesis, but each method has its own limitations. Although the CVD method can be used to produce high-quality rGO films, the high-temperature reaction (over 500 °C) may become an obstacle to industrialization. Liquid-phase exfoliation is a process that produces rGO nanosheets at room temperature under normal pressure, but this method can also lead to damage and uneven thickness of the nanosheets. Although wet chemical synthesis can control the surface morphology, grain size, and dopants of rGO materials, this method usually produces rGO materials of low quality and containing defects (Fig. 3.9) [34].

3.4.3 Innovative Preparation Technology

To develop new rGO film preparation technologies, an important trend in rGO research is to use green methods to synthesize high-quality rGO. This includes utilizing agro-industrial waste resources, resulting in significant environmental benefits. This green synthesis method is cost-effective and sustainable, and the resulting rGO has excellent morphology, uniform particle size, good optical properties, high conductivity, non-toxicity, and excellent chemical stability. Traditional synthetic methods, which often involve toxic solvents and produce harmful by-products, are being replaced by more environmentally friendly alternatives [35].

3.4.4 Performance Optimization

Further study and optimize the electrical, mechanical, and thermal properties of rGO films to meet the needs of specific applications. A modified Hummer method is used to produce GO, and then it is heated down in an argon atmosphere at 250 °C to obtain rGO. A series of steps in this process can be optimized, such as vigorous stirring, heating, addition of various chemicals, and vacuum filtration, to achieve the best quality films [3]. Defect engineering is used in graphene to create mid-gap states in the band structure, helping to slow down carrier recombination time and increase absorption efficiency thereby improving its conductivity and light detection properties [36, 37].

3.5 Summary

This chapter provides the preparation technologies of various commonly used rGO films and their application prospects in the fields of energy storage, sensors, and flexible electronic devices. It systematically outlines:

(i) A detailed discussion is conducted on the methods employed in the preparation of rGO films, including spin coating, vacuum filtration, spray coating, dip coating, drop casting, and electrophoresis.
(ii) The analysis highlights the broad application prospects of rGO films in the fields of energy storage, sensors, and flexible electronic materials.
(iii) RGO films present challenges in quality control and large-scale production, and future research needs to focus on developing new preparation techniques to improve the performance of rGO films and address existing challenges.

References

1. Zhang T et al (2024) Progress on the application of graphene-based composites toward energetic materials: a review. 31:95–116
2. Yang H et al (2017) Highly conductive free-standing reduced graphene oxide thin films for fast photoelectric devices. Carbon 115:561–570
3. Ahmed A et al (2023) Synthesis techniques and advances in sensing applications of reduced graphene oxide (rGO) Composites: a review. Compos A Appl Sci Manuf 165:107373
4. Chaabane L (2020) Graphene oxide-based materials: from water depollution to the catalysis of oganic reactions. Université de Lyon; Faculté des Sciences de Monastir (Tunisie)

References

5. Huang L et al (2018) Graphene-based nanomaterials for flexible and wearable supercapacitors. Small 14(43):1800879
6. Gascho JL et al (2019) Graphene oxide films obtained by vacuum filtration: X-ray diffraction evidence of crystalline reorganization. J Nanomater 2019:1–12
7. Songkeaw P et al (2019) Reduced graphene oxide thin film prepared by electrostatic spray deposition technique. Mater Chem Phys 226:302–308
8. Yaw CS et al (2020) Tuning of reduced graphene oxide thin film as an efficient electron conductive interlayer in a proven heterojunction photoanode for solar-driven photoelectrochemical water splitting. J Alloy Compd 817:152721
9. Zhang P et al (2023) A parameterized deposition rate model of electrostatic spraying rotating bell atomizer. 20(3):1019–1037
10. Sabzehmeidani MM, Gafari S, Kazemzad M (2024) Concepts, fabrication and applications of MOF thin films in optoelectronics: a review. 38:102153
11. Kim S-Y et al (2021) Microstructure and electrothermal characterization of transparent reduced graphene oxide thin films manufactured by spin-coating and thermal reduction. Results Phys 24:104107
12. Malik A et al (2024) Retrofitting of a full-scale dewatering operation for industrial polymer effluent sludge. 12(4):703
13. Fan X et al (2020) Hydrothermal reduced graphene oxide membranes for dyes removing. Sep Purif Technol 241:116730
14. Vakili S (2024) Development of an overall conversion concept from water to solvent-based paint system in the MAN Truck & Bus SE cab paint shop in Munich plant. Technische Hochschule Ingolstadt
15. Struchkov NS et al (2020) Uniform graphene oxide films fabrication via spray-coating for sensing application. Fullerenes, Nanotubes, Carbon Nanostruct 28(3):214–220
16. Fang M et al (2019) Preparation of highly conductive graphene-coated glass fibers by sol-gel and dip-coating method. J Mater Sci Technol 35(9):1989–1995
17. Shahzad S et al (2022) The effect of thin film fabrication techniques on the performance of rGO based NO_2 gas sensors at room temperature. Chemosensors 10(3):119
18. Ilickas M et al (2023) ZnO UV sensor photoresponse enhancement by coating method optimization. 14:100171
19. Cheng X et al (2023) Electrophoretic deposition of coatings for local delivery of therapeutic agents. 136:101111
20. Zhang H et al (2024) Electrophoretic deposition of single nanoparticles. 40(6):2783–2791
21. Gao Y et al (2021) Controllable preparations and anti-corrosion properties of reduced graphene oxide films by binder-free electrophoretic deposition. Appl Surf Sci 563:150295
22. Politano GG et al (2016) Physical investigation of electrophoretically deposited graphene oxide and reduced graphene oxide thin films. J Appl Phys 120(19)
23. Tamang S et al (2023) A concise review on GO, rGO and metal oxide/rGO composites: Fabrication and their supercapacitor and catalytic applications. 169588
24. Bo Z et al (2014) Green preparation of reduced graphene oxide for sensing and energy storage applications. Sci Rep 4(1):4684
25. Chen L, Liu D, Yang P (2019) Preparation of α-Fe_2O_3/rGO composites toward supercapacitor applications. RSC Adv 9(23):12793–12800
26. Mishra RK et al (2020) A novel RGO/N-RGO supercapacitor architecture for a wide voltage window, high energy density and long-life via voltage holding tests. Chem Commun 56(19):2893–2896
27. Wang M et al (2014) Large-area, conductive and flexible reduced graphene oxide (RGO) membrane fabricated by electrophoretic deposition (EPD). ACS Appl Mater Interfaces 6(3):1747–1753
28. Nurazzi NM et al (2021) The frontiers of functionalized graphene-based nanocomposites as chemical sensors. Nanotechnol Rev 10(1):330–369
29. Abid et al (2018) Reduced graphene oxide (rGO) based wideband optical sensor and the role of temperature, defect states and quantum efficiency. Sci Rep 8(1):3537

30. Sundriyal P, Bhattacharya S (2020) Textile-based supercapacitors for flexible and wearable electronic applications. Sci Rep 10(1):13259
31. Yu X et al (2017) Graphene-based smart materials. Nat Rev Mater 2(9):1–13
32. Rabchinskii MK et al (2018) Facile reduction of graphene oxide suspensions and films using glass wafers. Sci Rep 8(1):14154
33. Ahn SI et al (2015) Large and pristine films of reduced graphene oxide. Sci Rep 5(1):18799
34. Choi SH et al (2022) Large-scale synthesis of graphene and other 2D materials towards industrialization. Nat Commun 13(1):1484
35. Manikandan V, Lee NY (2023) Reduced graphene oxide: biofabrication and environmental applications. Chemosphere 311:136934
36. Li F et al (2021) Defect engineering in ultrathin SnSe nanosheets for high-performance optoelectronic applications. ACS Appl Mater Interfaces 13(28):33226–33236
37. Peng Z et al (2020) Strain engineering of 2D semiconductors and graphene: from strain fields to band-structure tuning and photonic applications. Light: Sci Appl 9(1):190

Chapter 4
Structural, Morphology, and Chemical Species Properties of Reduced Graphene Oxide

Abstract This chapter carried out a detailed analysis of the physical properties of redox oxide (rGO) films by using X-ray lines (XRD), scanning electron microscopy (SEM), and Raman spectroscopy. The research results show that the rGO thin film in the reduction process increases the size of the particles, decreases the distance between the particles, and shows good crystal structure and electrical conductivity. The SEM image shows the layered structure and surface shape of the rGO film, while the spectral analysis reveals the changes in the chemical composition of the rGO film, especially the changes in the defect density and the degree of oxidation. Through detailed comprehensive analysis, rGO provides theoretical support and practical guidance for applications in energy storage, sensors, and flexible electronic equipment.

Keywords Reduced graphene oxide films · X-ray diffraction · Scanning electron microscopy · Raman spectroscopy

4.1 Introduction

Reduced graphene oxide (rGO) films have attracted significant attention due to their high electrical conductivity, mechanical strength, and flexibility [1–3]. These properties make rGO films ideal for applications in supercapacitors, sensors, and flexible electronic devices [2, 4]. The optimization and detailed characterization of rGO films are crucial for understanding the structural transformations and lattice changes during the reduction process [5, 6]. However, despite the great progress in rGO research, it is still necessary to fully understand its physical characteristics through a detailed characterization to fully exploit its potential advantages [7].

This chapter aims to gain a deeper understanding of the structural transformation of rGO films by conducting detailed physical characterization using techniques such as XRD, SEM, and Raman spectroscopy. By examining the structure, morphology, and chemical properties of rGO films, this study will provide theoretical support and practical guidance for the potential applications of rGO films in various fields such as energy storage and flexible electronics.

4.2 Structure Analysis

The XRD technology is used to examine the crystal structure of inorganic and organic materials [8]. Through XRD analysis, the position of atoms or ions in the crystal structure can be revealed, thereby analyzing key information such as its structure. This principle is realized by analyzing the XRD pattern of the crystal, because each crystal will produce its own unique pattern [9]. By analyzing the width and position of the peaks in the XRD spectrum, it is possible to determine the size of the crystal particles and provide information about the internal stress state of the material [10]. The essence of XRD is that it can classify crystalline materials based on the orderly and repeated arrangement of atoms in the plane [11]. When the X-ray irradiates the crystal, it forms a unique diffraction pattern, thereby revealing the microscopic structural characteristics of the material [5, 12, 13].

The crystalline transformations inherent in the green synthesis process of rGO from GO utilizing grapefruit extract are elucidated by XRD analysis. For GO, the XRD pattern exhibits a pronounced diffraction peak at $2\theta = 13.29°$. This peak is indicative of a d-spacing of 6.66 Å, which is characteristic of oxygenated functionalities interspersed within the graphene layers. The associated Full Width at Half Maximum (FWHM) value is recorded at 0.78720, and the particle size is estimated to be 10.2 (Fig. 4.1) [13].

Upon reduction to rGO, there is a notable shift in the peak position to $2\theta = 26.50°$, reflecting a decrease in d-spacing to 3.36 Å [14]. This reduction is attributed to the removal of oxygen-containing groups, facilitating a denser stacking of the graphene sheets. Concurrently, the FWHM diminishes to 0.09804, indicating an enlargement in crystallite size to 83.3 nm [13].

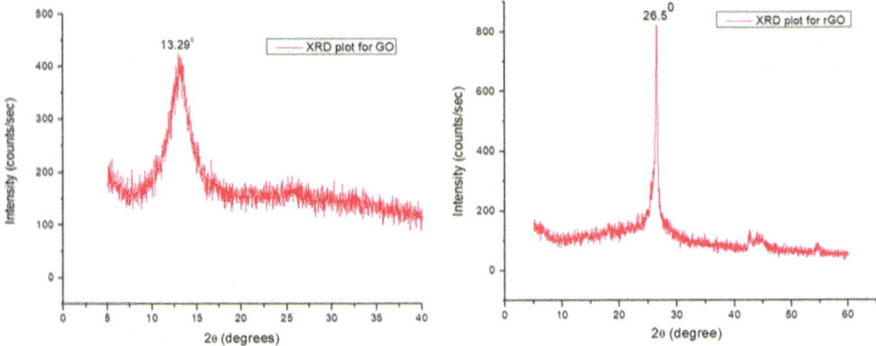

Fig. 4.1 XRD plots for GO and rGO, reproduced with permission from Ref. [13]

4.2 Structure Analysis

Compared to GO, rGO demonstrates lower oxygen content and shorter interlayer distance, which serves as sufficient evidence of the success of the reduction process. Furthermore, there is a marked increase in crystallite size and enhancement of the stacking order, rendering the rGO structure more analogous to pristine graphene [12]. The reestablishment of the sp^2 hybridized carbon network is expected to significantly bolster the electrical conductivity and mechanical robustness of rGO, which is significant for its applications in various fields [13, 15].

The improved Hummers method is used to synthesize GO from graphite powder, followed by thermal reduction of the synthesized GO at 250 °C for one hour under argon atmosphere to produce rGO [12]. The XRD spectra clearly reveal the structural changes from graphite to GO, and then to rGO, with each material exhibiting its unique characteristics (Fig. 4.2a) [12]. The graphite powder exhibits a pronounced diffraction peak at $2\theta = 26°$, reflecting its highly ordered layered structure. Upon oxidation to form GO, this peak shifts to a lower angle ($2\theta = 10°$) due to the introduction of oxygen and epoxy groups during the oxidation process, which expands the interlayer spacing (Fig. 4.2a) [12, 16].

The thermal reduction process converts GO into rGO, as evidenced by a strong diffraction peak at $2\theta = 25.1°$ in the XRD pattern. This peak indicates that the reduction process is accompanied by deoxygenation, gradually removing oxygen functional groups, reducing interlayer spacing, and restoring sp^2 bonding between carbon atoms. The resulting rGO exhibits a structure and properties similar to pristine graphene, highlighting its high electrical conductivity and structural integrity [12, 17].

Using the improved Hummers method, GO is derived from graphite, followed by reduction utilizing hydrazine hydrate as the reducing agent at 80 °C. The reduction process is meticulously analyzed by varying the reaction time from 5 to 120 min [5]. Examination of the XRD spectrum unveiled a prominent peak at $2\theta = 10.51°$, corresponding to an interlayer spacing of 0.841 nm. This measurement significantly

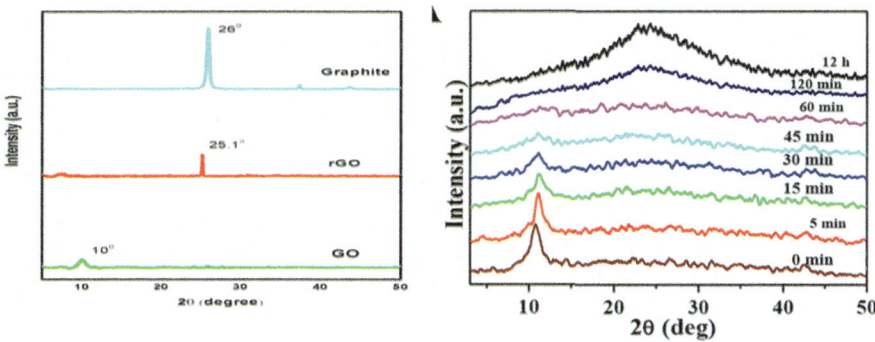

Fig. 4.2 a XRD patterns of graphite, GO, and rGO films, reproduced with permission from Ref. [12] and **b** XRD patterns of GO and rGO powder samples with different reduction times, reproduced with permission from Ref. [5]

exceeds that of natural graphite ($d = 0.334$ nm), primarily attributed to the incorporation of oxygen functional groups and intercalated water molecules, which disrupt the compact stacking of graphene layers [5, 18–20].

By observing the XRD spectrum, it is evident that at a reduction time of 12 h, the diffraction peak of rGO shifts to $2\theta = 23°$, indicating the removal of oxygen functional groups and the restoration of graphene-like structure [21]. This suggests that the reduction time is a key parameter in the reduction process of rGO, allowing for the adjustment of its performance in various application fields by controlling the reduction time (Fig. 4.2b) [5, 21, 22].

According to the XRD analysis results, the characteristic peak of graphite typically occurs at around $2\theta \approx 26.5°$, the characteristic peak of GO is located between 10–15°, while that of rGO is between 20 and 26.5°. The smaller angle of the characteristic peak for GO indicates the presence of oxygen functional groups and intercalated water molecules, resulting in a larger interlayer spacing. Conversely, rGO, achieved by the removal of oxygen functional groups from GO, restores a graphite-like structure, thereby exhibiting superior functionality and performance.

4.3 Morphologies Analysis

An advanced microscopy technique, SEM utilizes a high-energy electron beam to scan the surface of a sample, capturing signals emitted or scattered by electrons, thereby generating high-resolution images of the sample surface [23]. In the field of materials science, SEM is employed to comprehensively analyze the morphological features, layered structure, dimensional parameters, defect conditions, and impurity content of graphene samples, providing crucial information for further material characterization and performance evaluation [24].

Using an improved Hummers method, GO is synthesized from natural graphite. Subsequently, a 0.4 wt% GO aqueous dispersion is spin-coated onto a quartz substrate, with coating layers of 1 layer and 3 layers [23, 25]. The coated substrates are then placed in an argon atmosphere and subjected to high-temperature reduction ranging from 800 to 1000 °C to prepare rGO films.

Through high-resolution SEM image analysis of the rGO films, it is observed that both single-layer and three-layer rGO films exhibit relatively smooth surfaces with no apparent wrinkles or delamination [25, 26]. However, for the three-layer rGO film, a relatively rough surface is observed due to the partial accumulation or overlapping of graphene sheets. The uniform coating of rGO films on the quartz substrate is primarily attributed to the chemical interaction between GO particles and the substrate surface, known as chemical adsorption phenomenon (Fig. 4.3) [25].

Through the colloidal synthesis method, ZnO/rGO nanocomposite materials for nanofiltration membranes are prepared. High-resolution SEM analysis provided a deeper insight into the microstructure of the ZnO/rGO composite film. The film exhibited excellent flexibility and mechanical properties, even capable of undergoing 360° bending (Fig. 4.4a) [26]. Further SEM analysis revealed that the film is relatively

4.3 Morphologies Analysis

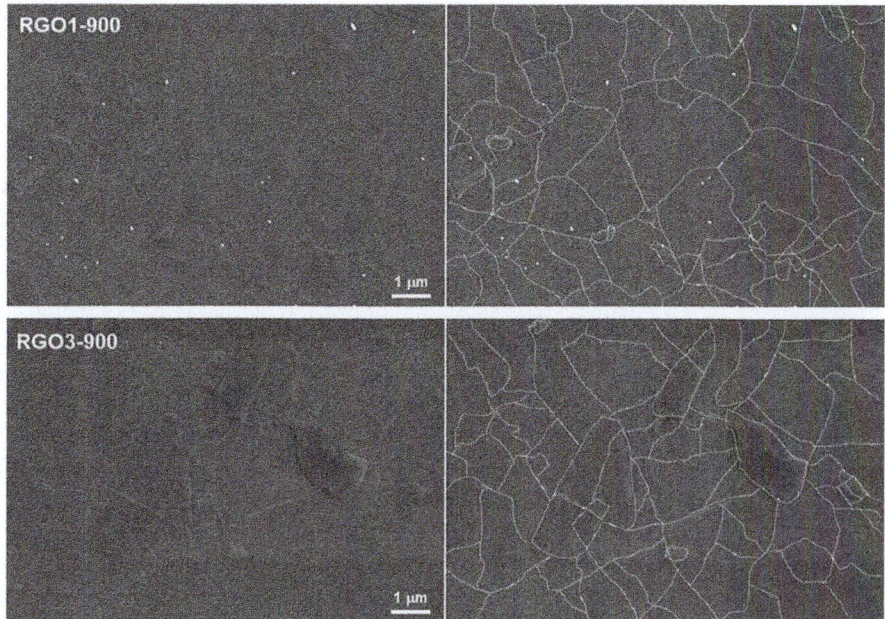

Fig. 4.3 SEM images of rGO thin films fabricated at 1 and 3 spin-coating layers and thermal reduction temperature of 900 °C. The white dotted lines in the image on the right represent the boundaries of the coated rGO platelets, reproduced with permission from Ref. [25]

smooth overall, without significant wrinkles or defects, indicating its high level of structural integrity (Fig. 4.4b) [25–27].

The cross-sectional SEM image reveals the layered structure of the ZnO/rGO film, which contributes to the enhancement of the material mechanical properties. During the synthesis process, the oxygen-containing functional groups on the GO surface undergo electrostatic attraction or chemical adsorption with Zn^{2+} ions, thereby facilitating the nucleation and growth of ZnO, ultimately forming the layered structure

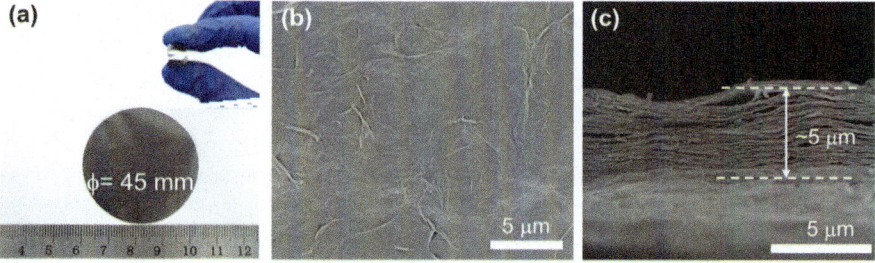

Fig. 4.4 a Digital photographs of the ZnO/rGO membranes, **b** top-view SEM image and **c** cross-sectional SEM image of ZnO/rGO membranes, reproduced with permission from Ref. [26]

of the ZnO/rGO composite material (Fig. 4.4c) [26]. This structure can significantly enhance the mechanical performance of the material.

Using a hydrothermal reduction method, the GO solution is hydrothermally reduced at 200 °C for 1–4 h [27–29]. Before the hydrothermal treatment, the GO sample displayed a layered structure, and dispersion in water yielded single-layer or few-layer GO (Fig. 4.5a) [27, 28]. TEM images of GO show that the thickness of single-layer GO film is 0.8–0.9 nm (Fig. 4.5e) [27]. After 1 h of reduction treatment, SEM images showed that the layered structure is preserved (Fig. 4.5b) [27], while TEM images indicated the coexistence of the layered structure of GO and the disordered structure of rGO, with a slight reduction in the number of layers compared to GO (Fig. 4.5f) [27].

When the reduction time is 2 h, the layered structure of GO becomes more disordered, exhibiting wrinkling and shrinking (Fig. 4.5c, g) [27]. With increasing hydrothermal time, these disordered and wrinkled features become more pronounced, indicating the gradual reduction and deoxygenation of GO (Fig. 4.5d, h) [27, 28]. This further demonstrates the effectiveness of hydrothermal treatment in promoting the transformation from GO to rGO.

Through further analysis using high-resolution SEM, the structural changes during the reduction process of GO to rGO can be clearly observed. During the reduction process, GO gradually evolves from its initial layered structure to a coexistence of layered GO structure and disordered rGO structure and ultimately transforms into a more disordered and wrinkled rGO structure. These structural changes are crucial for understanding the reduction mechanism from GO to rGO and its potential applications prospects.

4.4 Chemical Species Analysis

Raman spectroscopy is a scattering spectroscopy based on the interaction between light and materials. When light irradiates onto a substance, both elastic and inelastic scattering occur [30]. The scattered light in elastic scattering has the same wavelength as the excitation light, while in inelastic scattering, the scattered light includes components with wavelengths longer and shorter than the excitation light, collectively known as the Raman effect [31].

Raman spectra of rGO are typically measured using excitation wavelengths of 532 or 633 nm [31]. The Raman spectrum of rGO usually exhibits three key peaks: the D peak, G peak, and D' peak. The D peak occurs at around 1350 cm^{-1} and is primarily attributed to the symmetric stretching vibration breathing mode of sp^2 carbon atoms in the aromatic ring. The G peak appears at around 1580 cm^{-1} and is mainly caused by the stretching vibration between sp^2 carbon atoms. The G' peak is located at around 2680 cm^{-1} and is mainly due to the double phonon transition of carbon atoms with two opposite momenta [31, 32].

Raman spectroscopy provides detailed insights into the structure of GO and rGO, particularly in interpreting the peak at around 1600 cm^{-1} [33, 34]. Traditionally,

4.4 Chemical Species Analysis

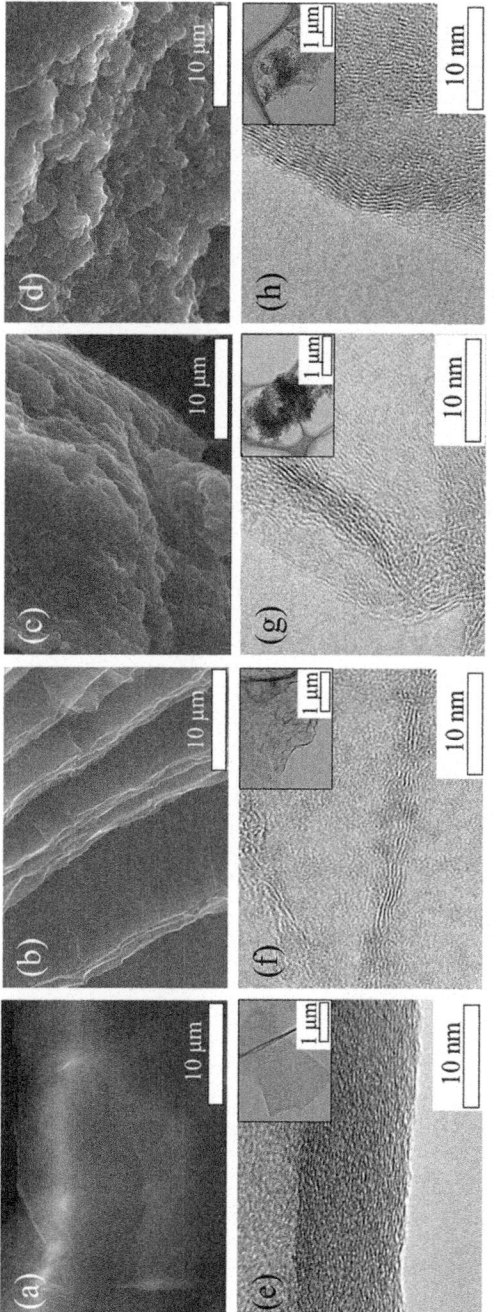

Fig. 4.5 SEM images of **a** GO, and its deoxygenated samples treated under **b** 1 h, **c** 4 h, and **d** 10 h. The corresponding TEM images are shown in (**e–h**), respectively. The insets of TEM images showing the sheets in low magnification, reproduced with permission from Ref. [27]

the G_{app} peak is considered as a single G peak, but it is actually composed of the superposition of the D and G' peaks. The D' peak arises from defects, and since both GO and rGO have high defect densities, the higher the intensity of the D' peak, the more significant its contribution to the G_{app} peak [33]. Now, by defining the G_{app} peak as arising from two modes and redefining the spectral differences between GO and rGO, a more accurate analysis is possible. During the reduction of GO to rGO, the G peak increases, while the intensity of the D' peak and the I_D/I_G ratio decreases, reflecting an increase in the degree of graphitization during the reduction process of rGO (Fig. 4.6) [31, 33].

Tetraethylammonium hydroxide (TEAH) is employed as both a reducing and stabilizing agent for the reduction of GO. Raman spectroscopy is employed to analyze the chemical structural changes [35]. The G band appeared at 1586 cm^{-1}, and the D band at 1319 cm^{-1}. The G band is associated with in-plane vibrations of sp^2-bonded carbon atoms, while the D band indicates the presence of structural defects. In the Raman spectra of rGO, the G band and D band appeared at 1583 cm^{-1} and 1328 cm^{-1}, respectively (Fig. 4.7) [35]. The ratio of I_D/I_G increased from 1.02 in GO to 1.17 in rGO, as observed in the spectral bands. These changes in band intensity and position indicate the reduction of GO induced by TEAH. The observed spectral changes signify the reduction process, characterized by a decrease in oxygen content and water molecules (Fig. 4.7) [31, 32, 35].

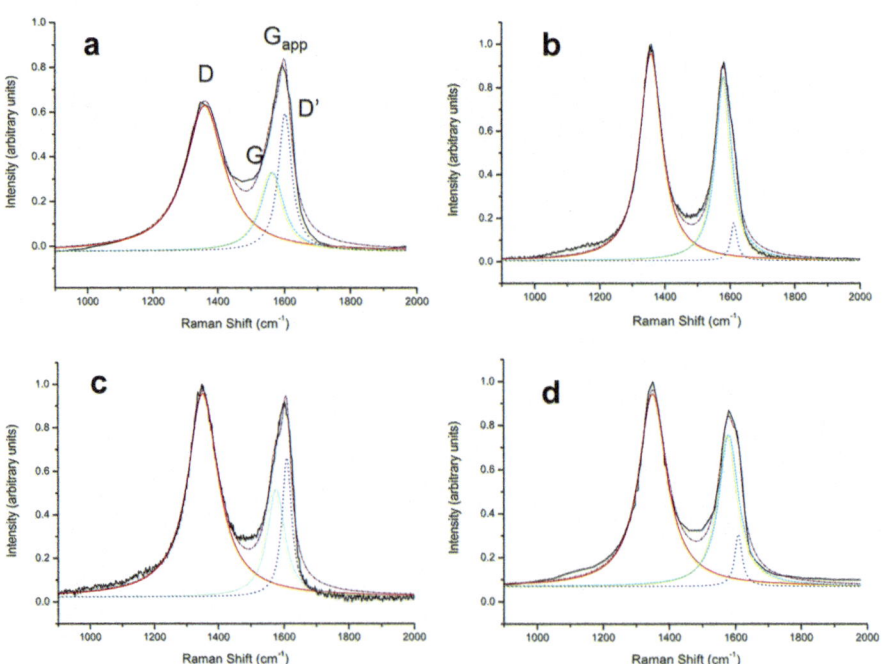

Fig. 4.6 **a, b** GO before and after thermal reduction at 1000 °C respectively and **c, d** GO before and after reduction with hydrazine, respectively, reproduced with permission from Ref. [33]

4.5 Summary

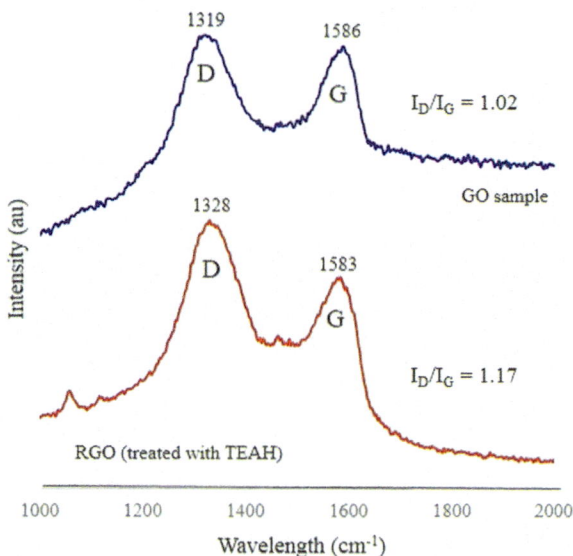

Fig. 4.7 Raman spectra of GO and rGO samples, reproduced with permission from Ref. [35]

The Raman spectra analysis of rGO and Au/rGO nanocomposite materials reveals that in rGO, the D band appears at 1360 cm^{-1}, which is associated with defects or vacancies in the carbon structure and the presence of sp^3 carbon. The G band appears at 1580 cm^{-1}, corresponding to the in-plane vibrations of sp^2-bonded carbon atoms. In the Au/rGO nanocomposite material, both the D (1350 cm^{-1}) and G (1575 cm^{-1}) bands shift to lower frequencies (Fig. 4.8) [36]. Additionally, the I_D/I_G intensity ratio of the Au/rGO nanocomposite material (0.81) significantly decreases compared to rGO (0.99), indicating that the introduction of Au nanoparticles helps reduce defects in the rGO sheets, thereby enhancing their quality.

Raman spectroscopy serves as a crucial tool in studying the process of GO to rGO conversion. By analyzing the D, G, and 2D peaks, insights into the structural changes during rGO reduction are gained, including an increase in the G peak, decrease in the D' peak, and changes in the I_D/I_G ratio. Spectral variations resulting from different reduction methods demonstrate the impact of the reduction process, while the introduction of Au nanoparticles aids in reducing defects in rGO, thereby enhancing its quality.

4.5 Summary

This chapter provides a detailed characterization of rGO, emphasizing its excellent physical properties and potential applications. It systematically outlines:

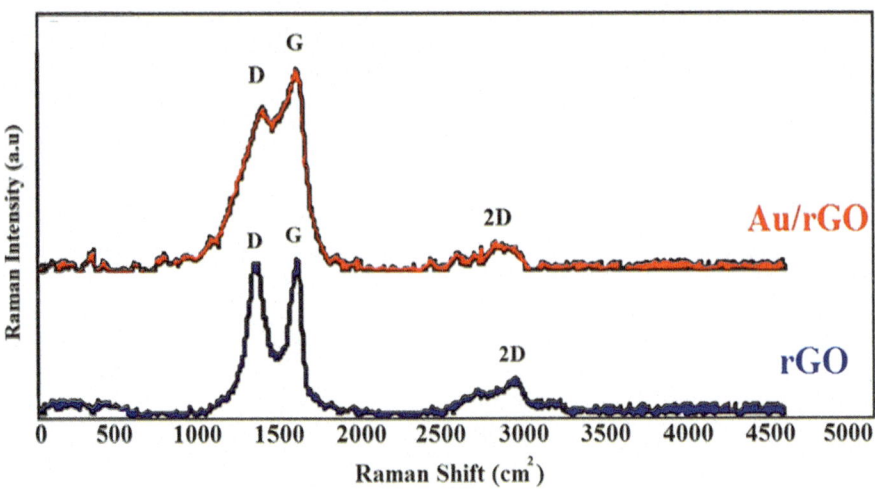

Fig. 4.8 Raman spectra of the prepared rGO, Au/rGO, reproduced with permission from Ref. [36]

(i) A detailed discussion is conducted on the characterization methods employed during the synthesis process of rGO, including XRD, SEM, and Raman spectroscopy.
(ii) Analyze the physical properties of rGO during the reduction process, including changes in crystal structure, surface morphology, and chemical composition.
(iii) Highlight the influence of the reduction process on the structural integrity, defect density, and graphitization level of rGO films, as well as their significance in the fields of energy storage and electronic materials.

References

1. Abdillah OB et al (2023) Recent progress on reduced graphene oxide and polypyrrole composites for high performance supercapacitors: a review. 74:109300
2. Ahmed A et al (2023) Synthesis techniques and advances in sensing applications of reduced graphene oxide (rGO) Composites: a review. 165:107373
3. Yao F et al (2024) Scalable one-step synthesis of reduced graphene oxide: towards flexible transparent conductive films and active supercapacitor electrodes. 488:150828
4. Huang J et al (2023) Rational design of electrode materials for advanced supercapacitors: from lab research to commercialization. 33(14):2213095
5. Das P et al (2024) Stepwise reduction of graphene oxide and studies on defect-controlled physical properties. 14(1):294
6. Nazer F et al (2024) Synthesis and characterization of black hydrogenated TiO_2-rGO nanocomposites with enhanced photocatalytic activity. 78:148–156
7. Wu J et al (2023) Graphene oxide for photonics, electronics and optoelectronics. 7(3):162–183
8. Jagiełło J et al (2020) Synthesis and characterization of graphene oxide and reduced graphene oxide composites with inorganic nanoparticles for biomedical applications. 10(9):1846

References

9. Taniguchi T et al (2023) Revisiting the two-dimensional structure and reduction process of graphene oxide with in-plane X-ray diffraction. 202:26–35
10. Ghorbani M et al (2023) Modified $BiFeO_3$/rGO nanocomposite by controlled synthesis to enhance adsorption and visible-light photocatalytic activity. 22:1250–1267
11. Lee XJ et al (2019) Review on graphene and its derivatives: synthesis methods and potential industrial implementation. J Taiwan Inst Chem Eng 98:163–180
12. Abid et al (2018) Reduced graphene oxide (rGO) based wideband optical sensor and the role of temperature, defect states and quantum efficiency. Sci Rep 8(1):3537
13. Gul W et al (2023) Synthesis of graphene oxide (GO) and reduced graphene oxide (rGO) and their application as nano-fillers to improve the physical and mechanical properties of medium density fiberboard. Front Mater 10:1206918
14. Singh SK et al (2023) Investigating the role of synthesized reduced graphene oxide and graphite micro-fillers on mechanical and fretting wear performance of glass fiber epoxy-based composite. 35(9):946–962
15. Alam SN, Sharma N, Kumar L (2017) Synthesis of graphene oxide (GO) by modified hummers method and its thermal reduction to obtain reduced graphene oxide (rGO). Graphene 6(1):1–18
16. Kashinath L et al (2017) Sol-gel assisted hydrothermal synthesis and characterization of hybrid ZnS-RGO nanocomposite for efficient photodegradation of dyes. 695:799–809
17. Dideikin AT, Vul' AY (2019) Graphene oxide and derivatives: the place in graphene family. Front Phys 6:149
18. Dawood AA, Moosa AA, Radhi MM (2022) Green synthesis of silver nanoparticles decorated with exfoliated graphite nanocomposites. 65(132):651–659
19. Li S et al (2022) Ultrahigh thermal and electric conductive graphite films prepared by g-C3N4 catalyzed graphitization of polyimide films. 430:132530
20. Xiao Z et al (2023) Spherical nano-graphite anode derived from electrochemical stripping for high performance Li-ion capacitors. 474:145623
21. Liu W, Speranza G (2021) Tuning the oxygen content of reduced graphene oxide and effects on its properties. 6(9):6195–6205
22. Sieradzka M et al (2021) The role of the oxidation and reduction parameters on the properties of the reduced graphene oxide. 11(2):166
23. Jaafar SMHSM (2017) Growth and characterization of graphene and graphene/copper oxide nanocomposites by hot-filament thermal chemical vapor deposition for flexible pressure sensor application. University of Malaya (Malaysia)
24. Schweizer P et al (2020) Low energy nano diffraction (LEND)–A versatile diffraction technique in SEM. Ultramicroscopy 213:112956
25. Kim S-Y et al (2021) Microstructure and electrothermal characterization of transparent reduced graphene oxide thin films manufactured by spin-coating and thermal reduction. 24:104107
26. Zhang W et al (2022) General synthesis of ultrafine metal oxide/reduced graphene oxide nanocomposites for ultrahigh-flux nanofiltration membrane. Nat Commun 13(1):471
27. Huang H-H et al (2018) Structural evolution of hydrothermally derived reduced graphene oxide. Sci Rep 8(1):6849
28. Ikram M et al (2020) Hydrothermal synthesis of silver decorated reduced graphene oxide (rGO) nanoflakes with effective photocatalytic activity for wastewater treatment. 15:1–11
29. Zhou X, Shi T, Zhou H (2012) Hydrothermal preparation of ZnO-reduced graphene oxide hybrid with high performance in photocatalytic degradation. 258(17):6204–6211
30. Cutroneo M et al (2019) Effects of the ion bombardment on the structure and composition of GO and rGO foils. 232:272–277
31. Lee AY et al (2021) Raman study of D* band in graphene oxide and its correlation with reduction. 536:147990
32. Korucu H et al (2023) The detailed characterization of graphene oxide. 77(10):5787–5806
33. King AA et al (2016) A New Raman Metric for the characterisation of graphene oxide and its derivatives. Sci Rep 6(1):19491
34. Muzyka R et al (2021) Characterization of graphite oxide and reduced graphene oxide obtained from different graphite precursors and oxidized by different methods using Raman spectroscopy statistical analysis. Materials 14(4):769

35. Cham sa-ard W et al (2021) Synthesis, characterisation and thermo-physical properties of highly stable graphene oxide-based aqueous nanofluids for potential low-temperature direct absorption solar applications. Sci Rep 11(1):16549
36. Yousefimehr F et al (2021) Facile fabricating of rGO and Au/rGO nanocomposites using Brassica oleracea var. gongylodes biomass for non-invasive approach in cancer therapy. Sci Rep 11(1):11900

Chapter 5
Cyclic Voltammetry Analysis of Reduced Graphene Oxide for Supercapacitors

Abstract Supercapacitors are known for their remarkable power density with rapid energy storage and release characteristics. This study utilized cyclic voltammetry (CV) to investigate the electrochemical properties of reduced graphene oxide (rGO) as an electrode material. The formation of the rGO electrode is achieved by reducing the graphene oxide (GO) and coating it onto a conductive substrate, followed by CV characterization carried out in KOH electrolyte using both 2-electrode and 3-electrode configurations. The outcomes demonstrated the rGO has better capacitance and charge storage than GO, with nearly rectangular CV curves further supporting its excellent electrochemical double-layer capacitance behavior. This chapter discusses the properties, efficiency, multifunctionality, and potential of rGO electrodes as supercapacitors by exploring their scan rate, reduction time, and composite materials.

Keywords Cyclic voltammetry · Reduced graphene oxide · Electric double-layer capacitor · Supercapacitors · Electrochemical performance analysis

5.1 Introduction

The investigation and exploration on materials selection for supercapacitors are getting essential to synchronize with the high demand of green and sustainable energy storage system [1, 2]. Reduced graphene oxide (rGO) gained significant interest from researchers due to its remarkable electrical conductivity, high specific surface area, and excellent electrochemical properties [3–5]. The reduction of rGO involves the sp^2 hybridization restoration of graphene oxide (GO) [4, 6]. Cyclic voltammetry (CV) is a fundamental electrochemistry characterization technique to characterize the capacitance behavior and energy storage capacity [7]. This chapter discusses the utilization of CV to analyze the properties of rGO as an ideal material for supercapacitor applications in details.

This chapter analyzes the effects of reduction time, scan rate and composite integration of rGO on supercapacitor performance. These analyses can help in identifying the optimal electrochemical performance, providing supporting evidences and guidelines for developing high-performance supercapacitors.

5.2 Experimental Setup for Cyclic Voltammetry

The basic principle of cyclic voltammetry is to apply a triangular waveform pulse voltage to a closed circuit formed by the working electrode and the counter electrode [6, 8, 9]. The potential at the working electrode/electrolyte interface is changed at a certain rate, forcing the active substances on the working electrode to undergo oxidation/reduction reactions [4, 10–12]. This process measures the response current during electrochemical reactions on the electrode [7, 13]. The electrode potential and the magnitude of the response current during this process are recorded to obtain the corresponding current–voltage curve. If the scanning voltage is only swept in one direction, from the starting potential Ui to the ending potential Us, the resulting current–voltage curve is called a linear voltammetric scanning curve (Fig. 5.1a) [7, 13–16].

The preparation of the WE involves using active material rGO, acetylene black, and polyvinylidene fluoride (PVDF) binder in a weight ratio of 8:1:1 [17–19]. An appropriate amount of N-methyl-2-pyrrolidone (NMP) solvent is added to the mixture, followed by thorough stirring and ultrasonic treatment to form a homogeneous slurry [20]. The slurry is then coated onto current collectors such as nickel foam or conductive glass [21]. The coating process can employ various techniques, such as spin coating or drop casting, depending on specific requirements. The coated electrodes are then dried and pressed to ensure that the rGO material adheres firmly to the current collector [22].

Experiments involving CV can be conducted using both standard two-electrode and three-electrode systems. In the two-electrode setup, a current collector coated with rGO serves as the working electrode (WE), opposed by a counter electrode

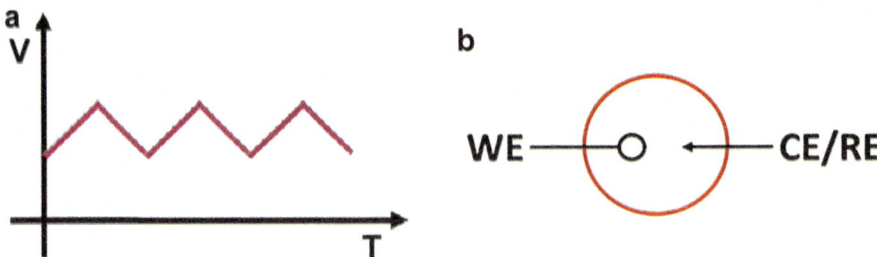

Fig. 5.1 **a** Pulse voltage of triangular wave and **b** a two-electrode cell

(CE), which can also be made of rGO, forming a symmetrical structure (Fig. 5.1b) [23].

On the other hand, the three-electrode system comprises the WE, CE, and reference electrode (RE). The WE is fabricated by depositing rGO onto a chosen current collector, while platinum wire typically serves as the CE and Ag/AgCl as the RE. The experiments are carried out using a 1 M KOH aqueous electrolyte solution, maintained at room temperature. CV measurements are conducted using a potentiostat, with parameters such as scan rate and voltage range adjusted based on the specific experimental requirements (Fig. 5.2a) [24]. The circuit and electrode reactions of a CV cell can comprehensively elucidate the electrochemical performance of an rGO electrode as it undergoes oxidation–reduction reactions with varying potentials, thereby providing a comprehensive understanding of its electrochemical properties (Fig. 5.2b, c) [24].

The three-electrode system consists of two circuits: one formed by the WE and the RE, which is used to test the electrode potential with very little current, and the other by the WE and the CE, which is used to test the current, the advantages of this system lie in its finer potential control and more accurate measurements, which are crucial for gaining a deeper understanding of the behavior of rGO in supercapacitor applications [25]. In the two-electrode system, the RE is eliminated, the RE circuit is combined with the WE circuit, the test current remains that of the WE circuit, and the voltage is the potential difference between the WE and the CE. This can better control the experimental conditions and help to explore the electrochemical properties of rGO more effectively [26].

In this section, the experimental setup for studying the electrochemical performance of rGO-based supercapacitors using CV is described in detail. The process is tested using a two-electrode and three-electrode system. The whole process includes preparation, coating, drying, and performance testing of the rGO working electrode. While the two-electrode system provides direct performance testing, the three-electrode system provides higher measurement accuracy and control capabilities to better explore potential applications of rGO in energy storage and conversion technologies.

5.3 Calculation for Cyclic Voltammetry

The specific capacitance (C) of the rGO film electrodes in the three-electrode system and two-electrode system is calculated from the cyclic voltammetry (CV) curves using the following Eqs. (5.1), (5.2), respectively [27].

$$C_m = \frac{1}{V \times m \times (U_2 - U_1)} \int_{U_1}^{U_2} I(U) dU \tag{5.1}$$

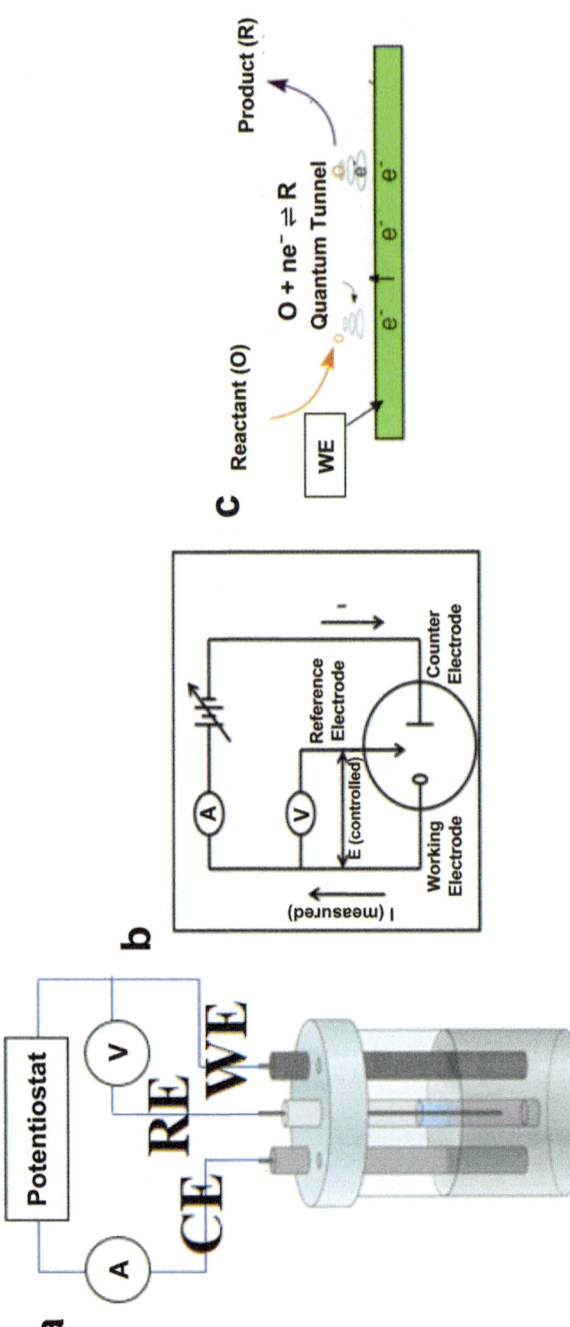

Fig. 5.2 **a** Three-electrode cell with the utilization of CV detection method—the input is the applied voltage while the output is the electrochemical analysis through the cyclic voltammogram curve, reproduced with permission from Ref. [24], **b** CV cell circuit—when the voltage applied to the electrochemical reaction starts on the electrode surface then the current starts to flow, reproduced with permission from Ref. [24] and **c** simulation of electrode reaction with the reactant—at a given applied voltage the reactant starts to move toward the electrode surface then the quantum tunneling happens to the electron; as a result, the reaction occurs to form the product, reproduced with permission from Ref. [24]

5.4 Cyclic Voltammetry of Graphene Oxide and Reduced Graphene Oxide

$$C_m = \frac{2}{V \times m \times (U_2 - U_1)} \int_{U_1}^{U_2} I(U) dU, \qquad (5.2)$$

where C_m (F g^{-1}), v(V S^{-1}), $U_2 - U_1$(V), and $U(V)$ refer to the mass-specific capacitance, scan rate of the CV curve, potential window, and discharge voltage, respectively.

5.4 Cyclic Voltammetry of Graphene Oxide and Reduced Graphene Oxide

In this section, by comparing with the CV curve of GO, the CV curve of rGO is closer to a rectangular shape and at the same scan rate, the integrated area enclosed by the CV curve of rGO is larger, indicating that rGO has better capacitance characteristics and charge storage capacity. Under ideal conditions, the CV curves of rGO are usually rectangular or nearly rectangular, which indicates that the rGO electrode material has excellent electrochemical double-layer capacitance behavior and fast charge and discharge capabilities [28–30]. If the CV curve is abnormal or asymmetric, it indicates a problem with the electrochemical performance, while peaks or obvious current differences in the curve indicate that the electrode material does not behave ideally within the operating voltage range [31].

The rGO curve displays a larger enclosed area than GO on the CV, indicating higher capacitance and more efficient charge storage capability (Fig. 5.3a) [32]. The quasi-rectangular shape of the rGO CV curve indicates its ideal capacitance behavior, which is essential for the fast charge–discharge cycle required in supercapacitors.

At the scanning rate of 100 mV s^{-1}, rGO and GO CV curves are compared and analyzed. The results show that the quasi-rectangle shape of rGO CV curves shows

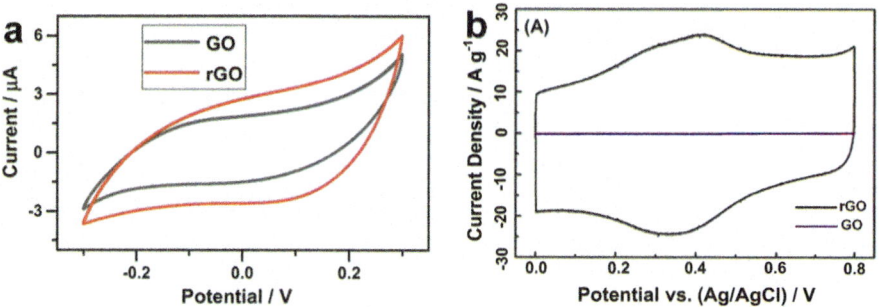

Fig. 5.3 **a** CV curves of GO and rGO at 20 mV s^{-1}, reproduced with permission from Ref. [32] and **b** CV curves of rGO and GO at scan rates of 100 mV s^{-1}, reproduced with permission from Ref. [33]

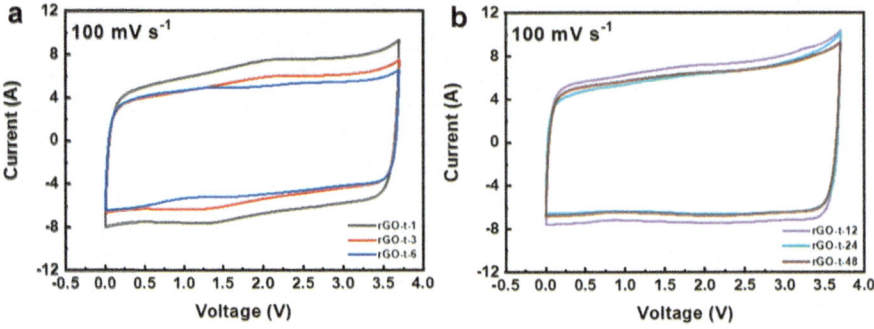

Fig. 5.4 a CV curves at voltage sweep rate of 100 mV s^{-1} of graphene synthesized with reduction time of 1, 3, 6 h and **b** 12, 24, 48 h, reproduced with permission from Ref. [34]

its ideal capacitance behavior. In addition, CV curves show no redox peaks in rGO, mainly because the main capacitive behavior of rGO is the double-layer capacitance (EDLC). In EDLC, the charge is stored in a double layer between the electrode surface and the electrolyte, rather than redox, suggesting that the reduction process enhances the stability and conductivity of rGO (Fig. 5.3b) [33].

The analysis of CV curves shows that the existence of oxygen-containing functional groups and reduction time are closely related to the electrochemical properties of rGO [34, 35]. The CV curve presents a nearly rectangular shape, showing the characteristics of an electrochemical double-layer capacitor. The samples rGO-t-1 and rGO-t-3 with shorter reduction times exhibit redox peaks due to oxygen-containing functional groups (Fig. 5.4a) [34].

The process as the reduction time increases from 1 to 48 h; these peaks decrease and disappear, indicating the further progression of the reduction process of rGO (Fig. 5.4a, b) [34, 36]. In addition, according to the area of the CV curve, it can be concluded that the electrochemical reaction related to the oxygen-containing functional group generates a small amount of pseudocapacitance [36].

This section demonstrates, through CV curve analysis, that rGO significantly surpasses GO in terms of capacitive performance and charge storage capacity. The CV curves of rGO exhibit behavior akin to an ideal capacitor, along with excellent electrochemical stability, ensuring outstanding rate performance and long-term cyclic usability under high power demands. These characteristics make rGO an ideal electrode material for supercapacitors.

5.5 Effect of Scan Rate on Cyclic Voltammetry

The scan rate in CV refers to the rate at which the potential of the WE changes over time. For the electrolyte ion adsorption state of an ideal supercapacitor, the shape of the CV curve will approach a rectangle. When the scan rate increases and causes

5.5 Effect of Scan Rate on Cyclic Voltammetry

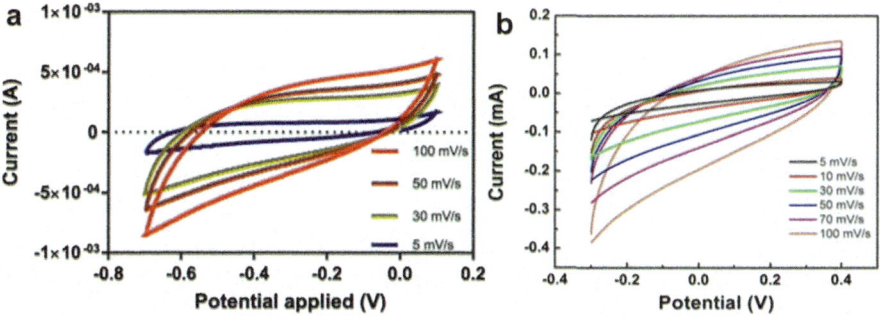

Fig. 5.5 **a** CV of rGO at different scan rates from 5 to 100 mV s^{-1}, reproduced with permission from Ref. [36] and **b** CV plot of rGO at varying scan rates from 5 to 100 mV s^{-1}, reproduced with permission from Ref. [5]

severe polarization of the electrolyte ions on the electrode, the shape of the CV curve will decay from a rectangle to a shuttle shape. Therefore, the scan rate is an important parameter for controlling the electrochemical response rate in the CV test [37]. In this section, the effect of different scan rates (from 5 to 500 mV/s) on the fast charge and discharge capabilities of rGO electrodes are investigated through CV analysis, which is crucial for optimizing its application in high-performance supercapacitors [27, 36].

From the CV curve of rGO, it can be observed that the area enclosed by the CV curve gradually increases as the scanning rate increases from 5 to 100 mV s^{-1}, indicating that with the increase of scanning rate, the area of the CV curve gradually increases, the charge storage also increases [5, 36, 38]. The shape of the curve also maintains similarity to the ideal capacitive rectangular shape, suggesting that the material exhibits excellent capacitive behavior, which is an important characteristic for fast charging and discharging cycles in energy storage applications (Fig. 5.5a, b) [5, 36].

In contrast analysis, the scan rate of rGO was increased from 5 to 500 mV s^{-1}. With the increase of scanning rate, the CV curve also shows the increase of current density [33]. This demonstrates not only the excellent performance of rGO as an electrode material for super capacitors in response to rapid electrochemical reactions, but also its ability to maintain high capacitance under different operating conditions (Fig. 5.6a, b) [27, 33].

By analyzing the performance of rGO at different scanning rates, the outstanding value of rGO in the application of super capacitor is highlighted. This indicates that rGO materials have excellent capacitance and adaptability to high-speed electrochemical processes. RGO, as an emerging material, has propelled the advancement of efficient energy storage systems and occupies a crucial position in the field of energy storage materials.

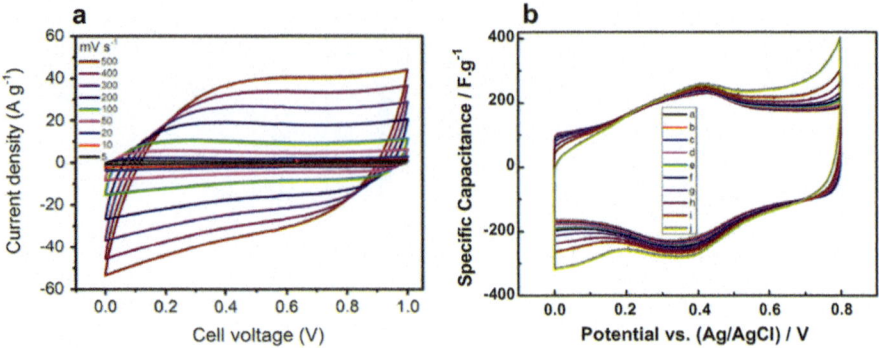

Fig. 5.6 a CV curves at different scan rates from 5 to 500 mV s^{-1}, reproduced with permission from Ref. [27] and **b** CV of rGO at different scan rates from 10 to 500 mV s^{-1} (from j to a), reproduced with permission from Ref. [33]

5.6 Effect of Composite Materials on Cyclic Voltammetry

Composite materials are new materials composed of two or more different materials, which possess superior properties compared to synthetic materials. In the process of advancing energy technology innovations, exploring new composite materials plays a crucial role in the development of supercapacitors with superior performance. This section focuses on the analysis of MoS$_2$/rGO prepared by composite molybdenum disulfide (MoS$_2$) and rGO nanomaterials as supercapacitor electrode materials [39, 40]. By using a three-electrode system for CV test analysis, the aim is to combine the advantages of these two materials, give play to their synergistic effect, and achieve an efficient electron transfer process, thereby further optimizing the design and performance analysis of high-performance supercapacitors [24].

The development of high-performance supercapacitors based on MoS$_2$ and rGO incorporated with NiO nanoparticles. The three-electrode system employed for measurements included rGO MoS$_2$ and MoS$_2$/rGO as working electrodes, a platinum rod as the counter electrode, and an Ag/AgCl electrode as the reference electrode during CV testing. Measurements are conducted at various scan rates (25, 50, 100, and 200 mV s^{-1}) within a voltage window of -0.6 to 0.6 V [39, 40].

The overall CV curve is mainly rectangular, indicating excellent capacitance performance even at high scan rates. The absence of redox peaks indicates that it is mainly contributed by the EDLC, which provides the electrode with reversible and rapid charge–discharge properties [38, 41]. It is worth noting that the current density of MoS$_2$/rGO electrode is significantly higher than that of pure MoS$_2$ electrode, which indicates that the introduction of rGO enhances the conductivity and activity of the electrode material and improves electron transfer efficiency (Fig. 5.7a, b) [40]. For further verification, MoS$_2$/rGO composites exhibit higher capacitance, lower internal resistance, and faster charge–discharge rate than pure MoS$_2$ electrodes; at the same scanning rate, it is shown that the introduction of rGO effectively

5.6 Effect of Composite Materials on Cyclic Voltammetry

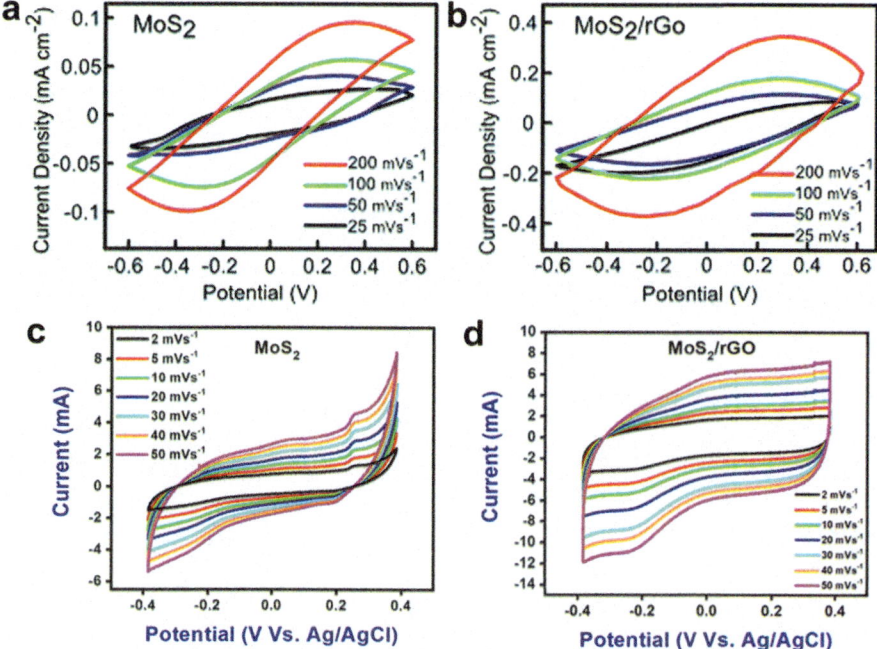

Fig. 5.7 **a** CV curves of MoS$_2$ electrode at different scan rates from 25 to 200 mV s^{-1}, reproduced with permission from Ref. [40], **b** CV curves of MoS$_2$/rGO electrode at different scan rates from 25 to 200 mV s^{-1}, reproduced with permission from Ref. [40], **c** CV curves of MoS$_2$ electrode at different scan rates from 2 to 50 mV s^{-1}, reproduced with permission from Ref. [39], and **d** CV curves of MoS$_2$/rGO electrode at different scan rates from 2 to 50 mV s^{-1}, reproduced with permission from Ref. [39]

improves the capacitance performance of the material (Fig. 5.7c, d) [5, 27, 33, 39, 40]. This study highlights the potential of adding rGO films to improve supercapacitor performance.

Integrating NiO nanoparticles into MoS$_2$ and rGO nanomaterials significantly enhances the performance of supercapacitors. Analysis of the CV curve in this study comprehensively demonstrates the synergistic effect among these nanomaterials, highlighting the superior performance of the MoS$_2$/rGO composite electrode over single-material counterparts. These findings not only deepen our understanding of the electrochemical behavior of composite material electrodes but also lay a solid foundation for the future development of high-performance supercapacitors.

5.7 Summary

This chapter provides a detailed analysis of rGO using CV, emphasizing its outstanding performance and potential in electrochemical applications. It systematically outlines:

(i) Through optimized rGO electrode preparation, scan rate control, and composite material integration, the efficiency of supercapacitors is improved.
(ii) The CV analysis highlights rGO nearly ideal electrostatic characteristics for application as an EDLC.
(iii) The rGO and its composite material electrodes exhibit higher capacitance, lower internal resistance, and faster charge–discharge rates, rendering them better suited for application in supercapacitors.

References

1. Das P et al (2024) Stepwise reduction of graphene oxide and studies on defect-controlled physical properties. 14(1):294
2. Muzaffar A et al (2024) Green supercapacitors: latest developments and perspectives in the pursuit of sustainability. 195:114324
3. Abdillah OB et al (2023) Recent progress on reduced graphene oxide and polypyrrole composites for high performance supercapacitors: a review. 74:109300
4. Ahmed A et al (2023) Synthesis techniques and advances in sensing applications of reduced graphene oxide (rGO) Composites: a review. 165:107373
5. Tamang S et al (2023) A concise review on GO, rGO and metal oxide/rGO composites: fabrication and their supercapacitor and catalytic applications. J Alloys Compounds 947:169588
6. Karim R et al (2024) Multiplying the water splitting performance of reduced graphene oxide–platinum nanoparticle hybrid by intercalating ethylenediamine polar space group
7. dos Santos JPA et al (2023) Best practices for electrochemical characterization of supercapacitors. 80:265–283
8. Houam S (2024) Theoretical and experimental study of the electrochemical detection of chemical compounds in aqueous medium
9. Schneider J, Tichter T, Roth C (2023) Electrochemical methods. In: Flow batteries: from fundamentals to applications, vol 1. pp 229–262
10. Garg N, Ramadesigan V, Tatiparti SSV (2023) Principles of Electrochemistry. In: Handbook of sodium-ion batteries. Jenny Stanford Publishing, pp 33–61
11. Ma M et al (2023) Towards stable electrode–electrolyte interphases: regulating solvation structures in electrolytes for rechargeable batteries. 2(6):833–854
12. Meena D et al (2023) Energy storage in the 21st century: a comprehensive review on factors enhancing the next-generation supercapacitor mechanisms. 72:109323
13. Rafiee M et al (2024) Cyclic voltammetry and chronoamperometry: mechanistic tools for organic electrosynthesis
14. Bisquert JJPE (2024) Inductive and capacitive hysteresis of current-voltage curves: unified structural dynamics in solar energy devices, memristors, ionic transistors, and bioelectronics. 3(1):011001
15. Smutok O, Katz E (2024) Electroanalytical instrumentation—how it all started: history of electrochemical instrumentation. 28(3):683–710

16. ul Haque S et al (2023) Electrochemistry analytical techniques and interpretation of the results. 1:53–76
17. Joseph A et al (2024) Recent advances in and perspectives on binder materials for supercapacitors—a review. 112941
18. Okhay O, Tkach A (2024) A comprehensive review of the use of porous graphene frameworks for various types of rechargeable lithium batteries. 80:110336
19. Salleh NA et al (2023) Electrode polymer binders for supercapacitor applications: a review. 23:3470–3491
20. Fernandez-Diaz L et al (2023) Mixing methods for solid state electrodes: techniques, fundamentals, recent advances, and perspectives. 464:142469
21. Tzeng Y et al (2023) Si–Ni-alloy-assisted very high-areal-capacity silicon-based anode on Ni foam for lithium ion battery. 30:101570
22. Huang J et al (2023) Rational design of electrode materials for advanced supercapacitors: from lab research to commercialization. 33(14):2213095
23. Xu J et al (2020) NiO-rGO composite for supercapacitor electrode. 18:100420
24. Fahmy Taha MH, Ashraf H, Caesarendra W (2020) A brief description of cyclic voltammetry transducer-based non-enzymatic glucose biosensor using synthesized graphene electrodes. Appl Syst Innovation 3(3):32
25. Kumar R, Thangappan R (2022) Electrode material based on reduced graphene oxide (rGO)/transition metal oxide composites for supercapacitor applications: a review. 5(6):1881–1897
26. Yang M et al (2024) RuNi single-atom alloy anchored on rGO as an outstanding bifunctional catalyst for efficient electrochemical water splitting
27. Zhu J et al (2020) Self-assembled reduced graphene oxide films with different thicknesses as high performance supercapacitor electrodes. J Energy Storage 32:101795
28. Aytug T et al (2018) Vacuum-assisted low-temperature synthesis of reduced graphene oxide thin-film electrodes for high-performance transparent and flexible all-solid-state supercapacitors. 10(13):11008–11017
29. Okhay O, Tkach AJN (2021) Graphene/reduced graphene oxide-carbon nanotubes composite electrodes: from capacitive to battery-type behaviour. 11(5):1240
30. Zhan Y et al (2020) Electrochemical deposition of highly porous reduced graphene oxide electrodes for Li-ion capacitors. 337:135861
31. Guo T et al (2022) Perspectives on working voltage of aqueous supercapacitors. 18(16):2106360
32. Dywili N et al (2023) High power asymmetric supercapacitor based on activated carbon/reduced graphene oxide electrode system. 35:105653
33. Yang J, Gunasekaran SJC (2013) Electrochemically reduced graphene oxide sheets for use in high performance supercapacitors. 51:36–44
34. Lin S et al (2023) Tuning oxygen-containing functional groups of graphene for supercapacitors with high stability. 5(4):1163–1171
35. Sanchez-Padilla N et al (2021) Influence of doping level on the electrocatalytic properties for oxygen reduction reaction of N-doped reduced graphene oxide. Int J Hydrogen Energy 46(51):26040–26052
36. Faiz MA et al (2020) Low cost and green approach in the reduction of graphene oxide (GO) using palm oil leaves extract for potential in industrial applications. Results Phys 16:102954
37. Malode SJ et al (2022) Preparation and performance of WO3/rGO modified carbon sensor for enhanced electrochemical detection of triclosan. 429:141010
38. Ramli NI et al (2018) Cyclic voltammetry and electrical impedance spectroscopy of electrodes modified with PEDOT: PSS-reduced graphene oxide composite. In: Transparent conducting films. IntechOpen
39. Baig MM et al (2020) High-performance supercapacitor electrode obtained by directly bonding 2D materials: hierarchal MoS2 on reduced graphene oxide. Front Mater 7:580424
40. Ghasemi F et al (2017) A high performance supercapacitor based on decoration of MoS 2/reduced graphene oxide with NiO nanoparticles. RSC Adv 7(83):52772–52781
41. Payami E et al (2020) Design and synthesis of novel binuclear ferrocenyl-intercalated graphene oxide and polyaniline nanocomposite for supercapacitor applications. 342:136078

Chapter 6
Galvanostatic Charge–Discharge Analysis of Reduced Graphene Oxide for Supercapacitors

Abstract Supercapacitors are extensively studied in the energy storage field because of their high power density and cycle stability compared to batteries. Conducting a thorough electrochemical analysis is essential for developing high-performance electrode materials. This chapter offers an in-depth examination of the electrochemical performance of reduced graphene oxide (rGO)-based electrodes using galvanostatic charge–discharge (GCD) testing to fully unlock the potential of rGO in supercapacitors. The findings indicate that rGO-based electrodes, thanks to their superior material properties, can achieve higher energy density and better stability. The combined effects of reduction time, current density, and nanocomposites on electrochemical performance are analyzed in detail, highlighting their significant potential to enhance the capacity and efficiency of energy storage systems.

Keywords Supercapacitors · Galvanostatic charge–discharge · Reduced graphene oxide · Energy storage · Electrochemical performance

6.1 Introduction

Supercapacitors are promising energy storage systems due to their high capacitance, excellent stability, rapid charge–discharge capabilities, and cost-effective maintenance [1–4]. However, the inherently low energy density of supercapacitors limits their development and applications; therefore, enhancing energy density and cycle life is crucial [5–7]. Researchers are interested in using reduced graphene oxide (rGO) as a carbon-based electrode because it improves electrical conductivity and cycle life while being cost-effective [8–10]. Understanding the electrochemical properties of rGO is essential to realize its full potential in supercapacitors [11–13]. Galvanostatic charge–discharge (GCD) is a key method for comprehensively understanding the specific capacitance, energy density, power density, and cycling stability of supercapacitors [14]. This chapter examines the role of GCD testing in evaluating rGO-based electrodes, focusing on:

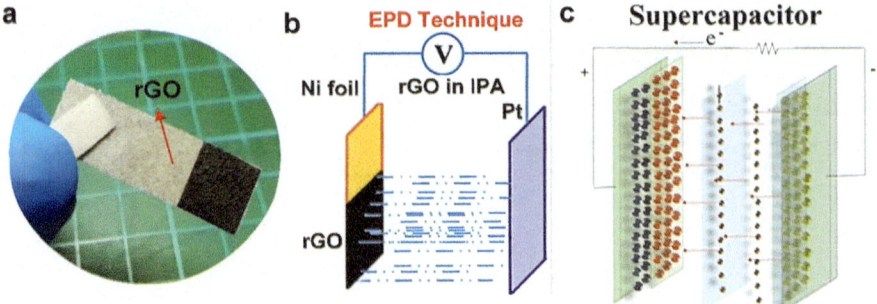

Fig. 6.1 a Using rGO as the electrode material for the working electrode, reproduced with permission from Ref. [20], **b** supercapacitors with rGO as the electrode material, and **c** rGO-based supercapacitor, reproduced with permission from Ref. [21]

(i) The detailed preparation of rGO electrodes and the configuration of the GCD system, emphasizing accurate measurement.
(ii) The introduction of various calculation formulas applicable to the GCD technique for analyzing multiple parameters.
(iii) The enhancement and optimization of supercapacitor performance through the study of reduction time, current density, and the integration of nanomaterials.

6.2 Charge–Discharge Fabrication for Reduced Graphene Oxide Electrodes

The preparation of rGO-based working electrodes (WE) begins with the preparation of a uniform rGO slurry and coating it on a current collector such as nickel foam and conductive glass [15–18]. The slurry is then coated onto nickel foam (1 cm × 1 cm), followed by overnight drying and pressing to serve as the WE (Fig. 6.1) [18–21]. Additionally, by adjusting the concentration of active materials, incorporating various dopants such as nitrogen and sulfur, and combining with other metal oxides, the specific capacitance and cycle stability of the electrode can be significantly enhanced [22–24].

6.3 Charge–Discharge Calculation

Curves from GCD tests provide a comprehensive analysis of multiple parameters in energy storage devices, encompassing specific capacitance (C) (Eq. 6.1), energy density (E) (Eq. 6.2), power density (P) (Eq. 6.3), and cycling stability. These parameters are derived from the following relationships [25].

6.4 Effect of Reduction Time on Galvanostatic Charge–Discharge

$$C = \frac{I \times \Delta t}{m \times \Delta V}, \quad (6.1)$$

where I (A) = Current, Δt (s) = Discharging time, m (g) = Mass of active material and ΔV (V) = Potential window.

$$E = \frac{1}{2} \times C \times \Delta V^2, \quad (6.2)$$

where E (Wh/kg) = Energy density, C = Specific capacitance, V = Potential window.

$$P = \frac{E}{2\Delta t}, \quad (6.3)$$

where P (W/kg) = Power density.

6.4 Effect of Reduction Time on Galvanostatic Charge–Discharge

One of the processes to improve electrochemical properties of rGO is reduction time. Reduction time refers to the duration of chemical, thermal, or other reduction processes to remove oxygen-containing groups and restore the conductive graphene structure from GO. This process not only reduces GO to rGO but also significantly affects the electrochemical performance of rGO as an electrode material in supercapacitors [26].

From the GCD curves, it can be observed that the GCD curve of the rGO electrode is closer to linear with minimal IR drop compared to GO, indicating a fast current–voltage response [27]. The specific capacitance of the rGO electrode at a current density of 0.75 A g^{-1} is 69.1 F g^{-1}, which is twice that of GO at the same current density [26, 28]. This capacitive behavior demonstrates that rGO has fast charge–discharge capabilities and a high ion diffusion rate (Fig. 6.2a) [26].

As the reduction time for rGO increases from 5 to 60 h, the shape of the GCD curves becomes more linear. This trend indicates that longer reduction times lead to improved electrical conductivity and ion transport performance in rGO [29, 30]. Extending the reduction time likely helps to further remove oxygen functional groups, thereby enhancing the capacitive properties of the material (Fig. 6.2b) [29].

GCD testing of graphene nanosheets with different reduction times shows that specific capacitance decreases with increasing reduction time. This decline may be due to the reduction in specific surface area and changes in oxygen and nitrogen functional groups. Graphene reduced for 30 min (G30m) exhibits a specific capacitance of up to 192 F g^{-1} at 0.1 A g^{-1}, while graphene reduced for 36 h (G36h) maintains a specific capacitance above 120 F g^{-1}, demonstrating high rate performance comparable to microwave-irradiated graphene (Fig. 6.2c) [31, 32].

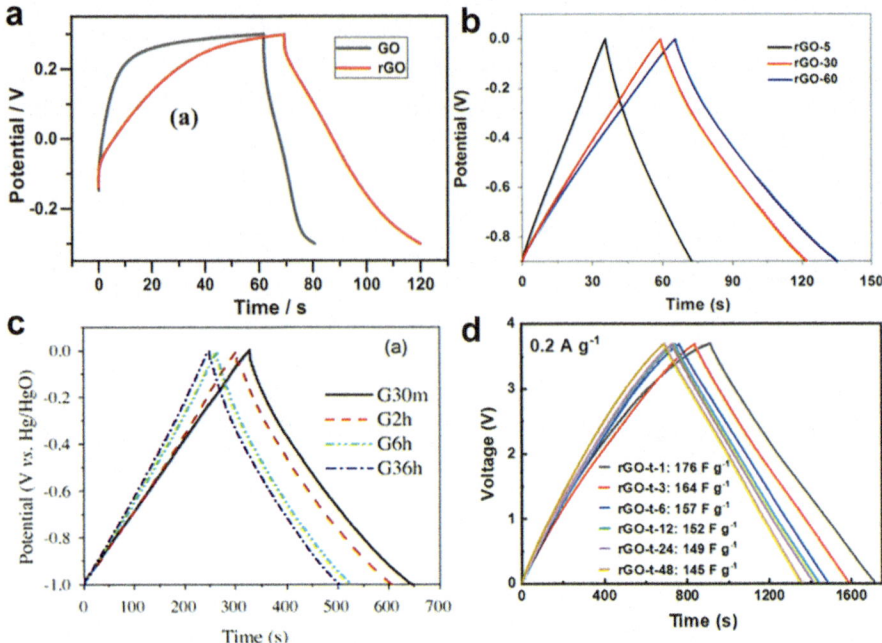

Fig. 6.2 **a** GCD curves of GO and rGO at 0.75 A g^{-1}, reproduced with permission from Ref. [26], **b** at a constant current density of 1 A g^{-1}, the GCD curves depict the supercapacitor performance of electrodes based on rGO-5, rGO-30, and rGO-60, reproduced with permission from Ref. [29], **c** the charge–discharge curves of graphene nanosheets at 0.5 A g^{-1}, reproduced with permission from Ref. [32], and **d** GCD curves of graphene reduced for 1–48 h, reproduced with permission from Ref. [36]

Samples of rGO treated with different reduction times exhibit a range of specific capacitance values, fully demonstrating the diversity and tunability of their electrochemical performance [33]. The variations in these specific capacitance values reveal an important fact: the structural and morphological characteristics of rGO change with different reduction times, and these changes, along with the degree of reduction, play a key role in determining its specific capacitance characteristics (Fig. 6.2d) [30, 34, 35].

The electrochemical performance of rGO as a supercapacitor electrode material is highly dependent on the preparation conditions, especially the reduction time. Optimal reduction conditions can find a balance between removing oxygen functional groups and maintaining structural integrity, thereby improving conductivity, ion transport, and consequently enhancing specific capacitance.

6.5 Effect of Current Density on Galvanostatic Charge–Discharge

In studying the electrochemical performance of various rGO materials, understanding their response to changes in current density during GCD cycles is crucial. Current density measures the electric current per unit cross-sectional area. In supercapacitors, it denotes the current passing through an electrode per unit area, typically expressed in amperes per square meter (A/m^2) [26]. Current density significantly affects the performance of supercapacitors [37]. At low current density, the measured mass change is less than that of the simple counter-ion adsorption process. At moderate current density, the mass change aligns with the counter-ion adsorption mechanism. At high current density, the additional mass change indicates the dense packing of ions and solvent molecules in the pores [38, 39]. This section delves into the detailed behaviors of rGO, nitrogen-doped rGO (N-rGO), and a composite of NiO with rGO (NiO/rGO), highlighting the influence of current density on their charge–discharge characteristics and overall performance, through systematic analysis of GCD curves at varying current.

The charge–discharge performance of rGO at varying current densities (0.1–0.5 A g^{-1}) reveals the impact of current density on its charge–discharge behavior. At lower current densities, the discharge time is significantly prolonged due to the enhanced diffusion of ions to the active sites within the electrode, thereby enabling more efficient charge storage. Conversely, higher current densities result in shorter discharge times, indicating limited ion accessibility and reduced utilization of the active sites. The nearly vertical charging curves indicate the rapid charging characteristics of the rGO material, while the more sloped discharge lines suggest higher internal resistance or diminished capacitive performance. This behavior highlights the dependence of electrochemical performance on current density and the impact of electrode resistance on charge–discharge efficiency (Fig. 6.3a) [40].

As the current density increases from 0.75 to 3.0 A g^{-1}, the charge–discharge curves of rGO become more leveled, reflecting the typical behavior of supercapacitors where the energy release rate of the capacitor is faster at higher discharge rates, hence the stored energy is depleted more quickly. There is no significant IR drop (a sharp drop in potential at the beginning of discharge) observed in any of the curves, indicating that the rGO material possesses good electrical conductivity (Fig. 6.3b) [26].

Similar to rGO, the GCD curves of N-rGO also show a decrease in charge–discharge time with increasing current density. However, notably at higher current densities, the slopes of the charge–discharge curves for N-rGO are more gradual, indicating that N-rGO has lower internal resistance and superior capacitive performance. Furthermore, the capacity retention ability of N-rGO decreases more slowly as the current density increases (Fig. 6.3a–c) [26, 41, 42].

Exploring the GCD behavior of NiO/rGO composite material within the current density range of 0.5–2.0 A g^{-1} reveals that the curves become increasingly nonlinear with the rise in current density, reflecting the pseudocapacitive behavior of NiO

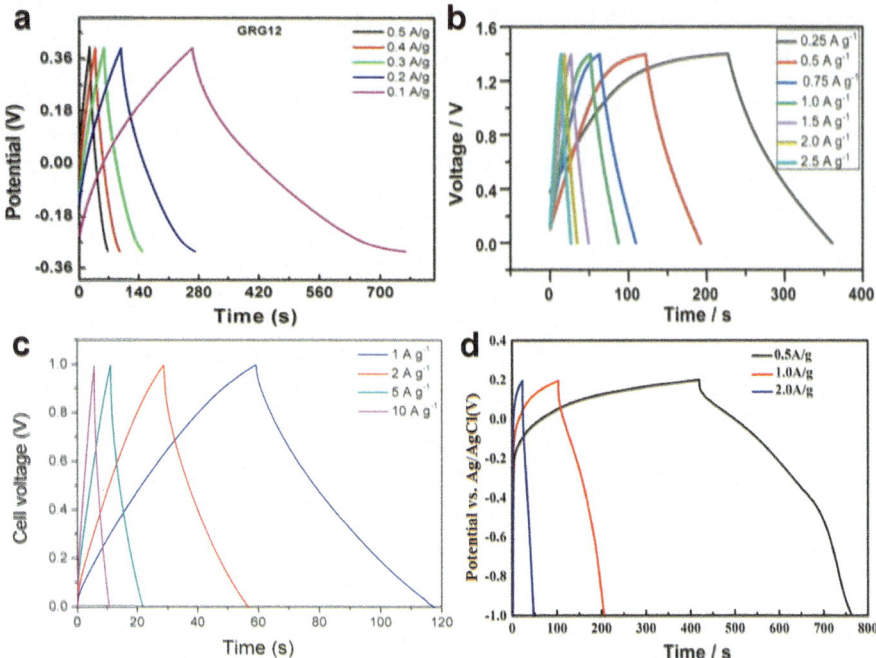

Fig. 6.3 **a** GCD curves of rGO at different current densities, reproduced with permission from Ref. [40], **b** GCD profiles of rGO at different current loads, reproduced with permission from Ref. [26], **c** GCD curves of N-rGO at different current densities, reproduced with permission from Ref. [42], and **d** GCD curves of NiO/rGO composite at a current density of 0.5 A g^{-1}, 1.0 A g^{-1}, 2.0 A g^{-1} respectively, reproduced with permission from Ref. [44]

due to Faradaic redox reactions [43]. Unlike the capacitive behavior of rGO and N-rGO, NiO/rGO provides higher specific capacitance at lower current densities, but this capacity decreases with an increase in current density, displaying typical characteristics of battery materials where the kinetics of redox reactions are limited at high rates (Fig. 6.3d) [44].

When comparing the electrochemical performance of rGO, N-rGO, and NiO/rGO composite materials, it is evident that nitrogen doping significantly enhances the charge–discharge efficiency and capacitance retention of rGO at high current densities. This improvement can be attributed to the increased electronic conductivity and additional active sites for charge storage introduced by nitrogen doping. Although the NiO/rGO composite material exhibits higher energy storage capacity at low current densities, it fails to maintain this advantage at high current densities.

6.6 Effect of Nanocomposite Materials on Galvanostatic Charge–Discharge

Nanocomposite materials enhance the overall performance of materials by combining nanoscale particles with a matrix material, leveraging the unique properties of nanoparticles. In the realm of supercapacitors, nanocomposite materials significantly improve the functionality of electrodes, offering breakthrough improvements in energy storage, energy density, and stability [45]. The interplay between these materials electrochemical behavior during constant current charge and discharge processes plays a key role in advancing energy storage technology [46].

This section analyzes the electrochemical properties of four nanocomposites: rGO/LaALO$_3$, MnO$_2$/rGO, rGO-ACP3, and MoTe$_2$/rGO. Through the GCD curve analysis of specific capacity, discharge potential, and capacitance, the unique synergistic effects of these composites relative to single components are highlighted, demonstrating their potential for application as high-performance supercapacitor electrode materials.

Comparison of the GCD curves of rGO/LaAlO$_3$ nanocomposite electrodes, pure rGO, and pure LaAlO$_3$ electrodes at different current densities. At the same current density, the discharge time of rGO/LaAlO$_3$ nanocomposite is longer than that of single rGO and LaAlO$_3$ electrodes, which indicates that the addition of rGO fully exerts the synergistic effect of the two pure materials and significantly improves their electrochemical performance (Fig. 6.4a) [45].

At a current density of 1 A g^{-1}, the charge–discharge curves of MnO$_2$, GO, and MnO$_2$/rGO nanocomposite electrodes indicate that the specific capacitance of the MnO$_2$/rGO nanocomposite electrode is 375 F g^{-1} (Fig. 6.4b) [46]. No redox peaks are observed in the charge–discharge curves, and their shape is consistent with the results of the CV curves [47]. This indicates that the MnO$_2$/rGO composite material has significant advantages in enhancing capacitive performance, particularly in improving specific capacitance and stability [46].

At a current density of 0.5 A g^{-1}, the comparison was made among activated carbon powder (ACP), rGO, and rGO-ACP3 regarding their GCD curves. From the curves, it is observed that the rGO-ACP3 electrode exhibits a significant IR drop, which may be due to higher internal resistance within the electrode. However, the discharge time of RGO-ACP3 is noticeably longer than that of the other materials, indicating it has higher capacitance (Fig. 6.4c) [47].

Exploring the GCD curves of MoTe$_2$, rGO, and MoTe$_2$/rGO at a current density of 0.5 A g^{-1}, the curves reveal that the MoTe$_2$/rGO electrode has the longest discharge time, indicating the highest specific capacitance and energy storage capacity. This composite material likely benefits from the synergistic interaction between the two components, optimizing the pathways for electron and ion transport (Fig. 6.4d) [48].

Nanocomposite materials, such as RGO/LaAlO$_3$ and MnO$_2$/RGO, often exhibit higher specific capacitance and energy density compared to single materials, due

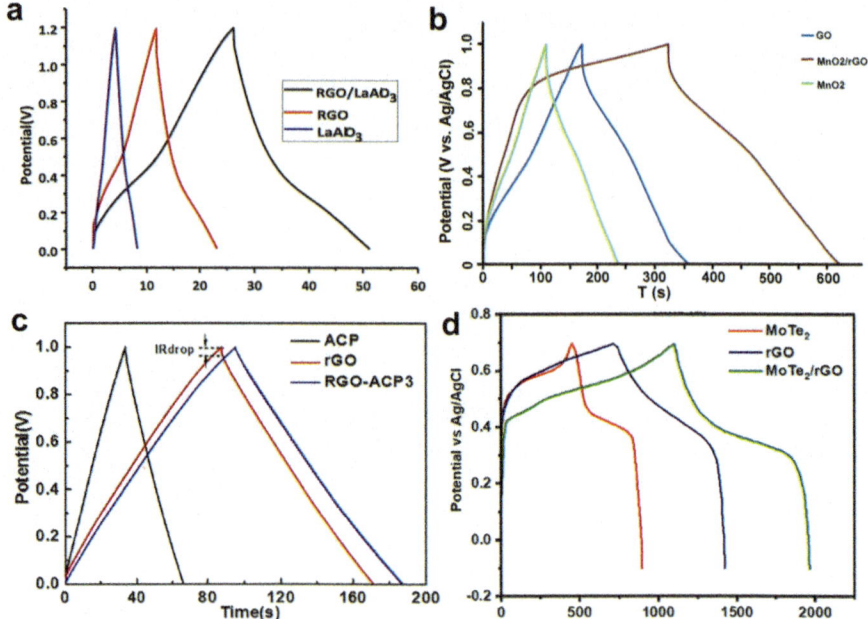

Fig. 6.4 **a** Comparison of GCD curves of RGO/LaAlO$_3$ nanocomposites, pure RGO and pure LaAlO$_3$ electrodes at constant current density of 5 mA, reproduced with permission from Ref. [45], **b** comparison of charge–discharge curves of MnO$_2$, GO and MnO$_2$/RGO electrodes at current density of 1 A g^{-1}, reproduced with permission from Ref. [46], **c** GCD curves at 0.5 A g^{-1} of ACP, Rgo, and RGO-ACP3, reproduced with permission from Ref. [47], and **d** comparison of GCD profile of different synthesized materials, reproduced with permission from Ref. [48]

to the synergistic effects and enhanced conductivity among the different components within the composite. Furthermore, composite materials contribute to improved cyclic stability of the electrodes, which is crucial for energy storage applications.

6.7 Summary

This chapter delves into the impact of GCD analysis on the electrochemical performance of rGO in supercapacitors. It systematically outlines:

(i) Advancements in supercapacitor efficiency through optimized rGO electrode preparation, current density control, and nanocomposite integration.
(ii) The rGO and its composites enhance capacitance, energy density, and stability, boosting energy storage.
(iii) The research enriches the understanding of the role of graphene in energy storage, setting the stage for future enhancements in supercapacitors and batteries.

References

1. Adedoja OS, Sadiku ER, Hamam YJP (2023) An overview of the emerging technologies and composite materials for supercapacitors in energy storage applications. 15(10):2272
2. Alam S et al (2023) Advancements in asymmetric supercapacitors: material selection, mechanisms, and breakthroughs with metallic oxides, sulfides, and phosphates. 72:108208
3. Khan HA et al (2024) A comprehensive review on supercapacitors: their promise to flexibility, high temperature, materials, design, and challenges. 131043
4. Phor L, Kumar A, Chahal SJ (2024) Electrode materials for supercapacitors: a comprehensive review of advancements and performance. 84:110698
5. Lu X et al (2023) Polymer-based solid-state electrolytes for high-energy-density lithium-ion batteries–review. 13(38):2301746
6. Wu W, Luo W, Huang YJCSR (2023) Less is more: a perspective on thinning lithium metal towards high-energy-density rechargeable lithium batteries. 52(8):2553–2572
7. Zhang H et al (2023) Prelithiation: a critical strategy towards practical application of high-energy-density batteries. 13(27):2300466
8. Ahmad F et al (2023) Advances in graphene-based electrode materials for high-performance supercapacitors: a review. 72:108731
9. Ahmed A et al (2023) Synthesis techniques and advances in sensing applications of reduced graphene oxide (rGO) Composites: a review. 165:107373
10. Waris et al (2023) A review on development of carbon-based nanomaterials for energy storage devices: opportunities and challenges. 37(24):19433–19460
11. Abdillah OB et al (2023) Recent progress on reduced graphene oxide and polypyrrole composites for high performance supercapacitors: a review. 74:109300
12. Shoeb M et al (2023) Unraveling the electrochemical properties and charge storage mechanisms of lactobacillus-mediated synthesized RGO-titanium silver nanocomposite as a promising binder-free electrode for asymmetric supercapacitor device. 964:171188
13. Umar A et al (2024) Exploring the potential of reduced graphene oxide/polyaniline (rGO@PANI) nanocomposites for high-performance supercapacitor application. 479:143743
14. Liu Y et al (2020) Activation-free supercapacitor electrode based on surface-modified Sr2CoMo1-xNixO6-δ perovskite. 390:124645
15. Abdullah T et al (2023) Engineering energy storage properties of rGO based Fe_2O_3/CuO/PANI quaternary nanohybrid as an ideal electroactive material for hybrid supercapacitor application. 299:117472
16. Ahmad F Utilization of $Sr_2Ni_2O_5$/rGO as electrode material in supercapacitor
17. Alsaiari M et al (2024) Effect of MoS2 and electrolyte temperature on the electrochemical performance of NiCoS@ rGO-based electrode material for energy storage, oxygen reduction reaction and electrochemical glucose sensor. 35:101909
18. Zeshan M et al (2024) Fabrication of niobium selenide-based rGO hybrid nanoparticles by hydrothermal method for supercapacitor applications. 50(4):7110–7120
19. Amiri M, Zardkhoshoui AM, Davarani SSHJN (2023) Fabrication of nanosheet-assembled hollow copper–nickel phosphide spheres embedded in reduced graphene oxide texture for hybrid supercapacitors. 15(6):2806–2819
20. Kumar A et al (2019) Synthesis of free-standing flexible rGO/MWCNT films for symmetric supercapacitor application. Nanoscale Res Lett 14:1–17
21. Tamang S et al (2023) A concise review on GO, rGO and metal oxide/rGO composites: fabrication and their supercapacitor and catalytic applications. 169588
22. Anwar MI et al (2023) Nitrogenous MOFs and their composites as high-performance electrode material for supercapacitors: recent advances and perspectives. 478:214967
23. Huang Z et al (2023) Stabilizing sulfur doped manganese oxide active sites with phosphorus doped hierarchical nested square carbon for efficient asymmetric supercapacitor. 468:143574
24. Kumar K, Kundu R (2024) Doping engineering in electrode material for boosting the performance of sodium ion batteries

25. Sajjad M et al (2021) Phosphine-based porous organic polymer/rGO aerogel composites for high-performance asymmetric supercapacitor. ACS Appl Energy Mater 4(1):828–838
26. Dywili N et al (2023) High power asymmetric supercapacitor based on activated carbon/reduced graphene oxide electrode system. Mater Today Commun 35:105653
27. Pratheepa MI, Lawrence M (2020) Eco-friendly approach in supercapacitor application: CuZnCdO nanosphere decorated in reduced graphene oxide nanosheets. SN Appl Sci 2:1–12
28. Okhay O, Tkach A (2021) Graphene/reduced graphene oxide-carbon nanotubes composite electrodes: from capacitive to battery-type behaviour. Nanomaterials 11(5):1240
29. Li Y-F et al (2015) Green synthesis of reduced graphene oxide paper using Zn powder for supercapacitors. Mater Lett 157:273–276
30. Xiao W et al (2023) Three dimensional graphene composites: preparation, morphology and their multi-functional applications. Compos A Appl Sci Manuf 165:107335
31. Bai J, Hong W, Bai H (2022) Electrochemically reduced graphene oxide: preparation, composites, and applications. Carbon 191:301–332
32. Fan L-Z et al (2012) The effect of reduction time on the surface functional groups and supercapacitive performance of graphene nanosheets. Carbon 50(10):3724–3730
33. Jabeen S, Kumar P, Samra KS (2024) Optimizing electrochemical performance: investigating the influence of oxidation of graphene oxide in rGO@ $MnMoO_4$. J Alloys Compound 173673
34. Arias Arias F et al (2020) The adsorption of methylene blue on eco-friendly reduced graphene oxide. Nanomaterials 10(4):681
35. Lesiak B et al (2021) Chemical and structural properties of reduced graphene oxide—dependence on the reducing agent. J Mater Sci 56:3738–3754
36. Lin S et al (2023) Tuning oxygen-containing functional groups of graphene for supercapacitors with high stability. Nanoscale Adv 5(4):1163–1171
37. Dong W et al (2023) Materials design and preparation for high energy density and high power density electrochemical supercapacitors. 152:100713
38. Sun K et al (2023) Electrocapacitive deionization: mechanisms, electrodes, and cell designs. 33(18):2213578
39. Zhang H et al (2023) Influence of ion exchange membrane arrangement on dual-channel flow electrode capacitive deionization: theoretical analysis and experimentations. 548:116288
40. Rai S et al (2021) Biocompatible synthesis of rGO from ginger extract as a green reducing agent and its supercapacitor application. 44(1):40
41. Das TK et al (2019) Electrochemical performance of hydrothermally synthesized N-Doped reduced graphene oxide electrodes for supercapacitor application. Solid State Sci 96:105952
42. Li S-M et al (2015) N-doped structures and surface functional groups of reduced graphene oxide and their effect on the electrochemical performance of supercapacitor with organic electrolyte. J Power Sources 278:218–229
43. Yousefipour K, Sarraf-Mamoory R, Yourdkhani A (2022) Supercapacitive performance of Fe-doped nickel molybdate/rGO hybrids: the effect of rGO. Colloids Surf, A 647:129066
44. Xu J et al (2020) NiO-rGO composite for supercapacitor electrode. 18:100420
45. Raj TV et al (2020) Facile synthesis of perovskite lanthanum aluminate and its green reduced graphene oxide composite for high performance supercapacitors. J Electroanal Chem 858:113830
46. Ghasemi S, Hosseini SR, Boore-Talari O (2018) Sonochemical assisted synthesis MnO_2/RGO nanohybrid as effective electrode material for supercapacitor. Ultrason Sonochem 40:675–685
47. Wang J et al (2020) To increase electrochemical performance of electrode material by attaching activated carbon particles on reduced graphene oxide sheets for supercapacitor. J Power Sources 450:227611
48. Abdullah M et al (2023) Development of binder-free $MoTe_2$/rGO electrode via hydrothermal route for supercapacitor application. Electrochim Acta 466:143020

The manufacturer's authorised representative in the EU is Springer Nature Customer Service Centre GmbH, Europaplatz 3, 69115 Heidelberg, Germany. If you have any concerns regarding our products, please contact ProductSafety@springernature.com

Printed and bound by CPI Group (UK) Ltd, Croydon, CR0 4YY

26/03/2026

02078940-0009